A Field Guide to Western Mushrooms

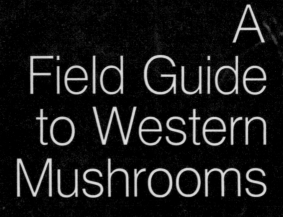

A
Field Guide
to Western
Mushrooms

by Alexander H. Smith

The University of Michigan Press
Ann Arbor

Copyright © by The University of Michigan 1975
All rights reserved
ISBN   0–472–85599–9
Library of Congress Catalog Card No. 74–25949
Published in the United States of America by
The University of Michigan Press and simultaneously
in Don Mills, Canada, by Longman Canada Limited
Manufactured in the United States of America

Sci
R

Contents

Introduction

The need for regional mushroom guides becomes increasingly acute each year as popular interest in the fleshy fungi grows. Mushroom hunting for whatever reason the hunter may have — a walk in the woods, mushrooms to eat, fungi to photograph, or specimens to paint — has taken its place with hunting mineral specimens, birdwatching, fishing, wild flower identification, and other educational types of out-of-door recreation. For years we have had regional vascular-plant floras, regional bird books, special books on minerals and geology, and books on wild flowers. But popular books on mushrooms for any region in North America are next to nonexistent. In fact popular mushroom books, by and large, contain a high percentage of the common mushroom "weeds" found generally throughout the world, and hence contain a great deal of duplication. It seems that two factors have operated to bring this about: first, publishers of mushroom books want the largest possible number of potential customers, so world coverage is attempted. Second, authors make an effort to include species that everyone is likely to find. In addition, most authors, and plant systematists as a group, have failed to appreciate the extent of the diversity among the higher fungi, with the result that it has been impossible to bring into focus, for interested parties, the true scope and nature of the problem of the recognition of species. Our knowledge of these fungi has increased rapidly only recently, and for the most part this information is still generally unavailable to amateur and many professional mycologists alike.

It is with the thought of taking a small step toward correcting this situation that this book on selected species of western fungi is offered. The number of species is small, 201 to be exact, but it offers more to the western mushroom hunter on his home region than any other comparably priced publication. Seventy of the species included here were known only from the western area or were described from it originally but have since been found elsewhere. Over twenty-five poisonous or very undesirable species for the table from the area are illustrated and described. Many species are in the "edible" column meaning that as far as it is known they are not poisonous. About fifteen of these are considered to be of "gourmet" calibre. Finally there is the residum of species of which the edibility is still apparently unrecorded. These have for the most part been listed as "not recommended." A work containing *all* the species in the area, by present knowledge and costs, would constitute several volumes and the price would be prohibitive. I am, in fact, in the process of writing such a set, but the project has taken a lifetime. It was started in 1935 and is still in progress. I have made eighteen extended expeditions to the West during spring, summer, and fall, but mostly during the latter when the major fruiting takes place. These expeditions have yielded over 50,000 collections. I was fortunate to be able to collect mushrooms in virgin stands of many species of forest tree before these were destroyed by cutting, and also have now collected in the extensive stands of various types of "scrub" vegetation which has come up following the cutting on both private and National Forest land. In other words, I have followed, to some extent, the ecological changes taking place with the wholesale destruction of habitats and their general revegetation. The present book is based on this experience.

Duplication with general popular books has been held to a minimum and occurs only for highly prized edible species, those which are very dangerous, or species where it was desired to bring out some feature not previously adequately emphasized. As far as possible an attempt has been made to include fungi with some exceptionally interesting feature, such as the occurrence of *Volvariella surrecta* on *Clitocybe nebularis* and *Collybia racemosa* with secondary branches on the stalk. Many species known only from our western area are included since these are the ones frequently encountered in the fields and woods, but not in the general literature. Such accounts as do exist are usually scattered in the professional literature, and thus are not available except in the best reference libraries. Lastly, but equally as important as any other consideration, species of particular interest to the mushroom hunter from the standpoint of edibility, but not generally treated in the literature of the area, have been emphasized, such as *Amanita phalloides* and *Boletus aereus.* It is interesting to note that both of the above species have assumed prominence in our western area during the past five years.

The Mushroom

So much has been written about the difficulties involved in the study of mushrooms and other fleshy fungi that many people are hesitant about pursuing it. However, mushrooms are not as complicated structurally as wild flowers. If one keeps certain principles in mind and is willing to follow certain rules, rapid and satisfactory progress can be made as teaching classes in field identification of fleshy fungi has repeatedly demonstrated. The would-be mushroom hunter must resign himself to becoming an opportunist; for the game of mushroom hunting is to do the most you can with what you happen to find. The next most important task is to learn how to look at a mushroom to get the most information possible from it. The time to do this is when you remove it from the soil or wood on which it is growing.

It must be kept in mind that the mushroom which is the object of the hunt is merely the fruiting stage of a plant (made up of threads or hyphae which grow *in* the soil or wood). The fruit of the mushroom plant (the mushroom) decays readily so that it is likely to be available for a limited time only. The length of this period of availability varies with the species from a few days to several weeks. Secondly, there are several thousand different kinds of fleshy fungi in the western region and not all of these fruit every season. Like wild flowers, some appear in the spring, summer, fall, or even winter in more southern states. In short, one can return to the same place year after year and each time find species he has never seen before.

It is the uncertainty of what one will find that adds zest to the sport of mushroom hunting. This aspect of the "game" has an important bearing on the question of which species are considered rare or common in an area. To the unimaginative hunter just beginning, all are rare. The experienced collector knows where to go for certain species at the right time, and to him the species he seeks are common in terms of his success. In the technical literature many species are listed as rare simply because they are generally unknown and go unrecognized when found. Others are rare in the sense that they are available for rather short periods of time. Others are rare because as a rule only one fruit body occurs at a time in a given place. Other reports of rarity simply mean that the collector has not been lucky enough to find the species in question.

Particularly in the West, the "lay of the land," the amount of rainfall, the months in which it occurs, and day as well as night temperatures (minimum and maximum), all have a bearing on the fruiting of mushrooms and related fungi. In nature, of course, all these factors operate in unison. But once it is recognized that uncertainty actually is a contributing factor to the pleasure of the hunt, you have made a good start. It is not necessary for me to lecture on the value of record keeping relative to your experiences.

One of the most important features of mushroom hunting, however, is to realize that mushrooms are selective, to a degree, in where they grow. To be a good hunter one should learn to recognize the different kinds of forest trees as there is often a direct relationship between a given species of tree and the presence of a particular kind of mushroom.

The most important factor in learning about mushrooms, after you have found some, is to observe critically all that can be seen on the fruit body at the time you pick it. The following discussion is intended as a guide to *how to look at a mushroom,* and what is it that is important for one to see.

The mushroom fruit body is a very simple structure in most species. It consists of a cap, a stalk, and on the underside of the cap, a spore bearing surface in the form of plates (termed lamellae or gills), narrow tubes or spines, or the surface may be flat to wrinkled. The technical term applied to this surface regardless of the form in which it occurs is hymenophore (bearing the hymenium, i.e., the spore bearing surface). The cap (pileus) is formed by expansion, by radial growth, of the apex of the stalk (stipe). The stalk is merely a mechanical device for lifting up the hymenophore so that the spores produced can be discharged and carried away by air currents. One will soon note that a stalk is lacking in many species — especially those occurring on wood. This appears to be the situation where a structure of the fruit body is no longer needed to aid in dispersal of spores, since the piece of wood itself serves essentially the same function as the stalk. However, not all wood inhabiting mushrooms have lost their stalks, so the question of "use and disuse" is still open for discussion. The stalk, when present, also serves as the channel through which the food materials necessary for spore production are transported from the vegetative plant to the hymenium of the fruit body.

The unit of structure of the fruit body is a thread called a hypha (pl. hyphae). In the stalk of most species the hyphae are arranged much in the manner of a series of pipes (in vertical arrangement) and bound together by other slender crooked threads called binding hyphae. This type of structure is often most evident in species with slender stalks. The stalk often extends deep into the ground — but of course this statement gives a false impression since the stalk develops from the vegetative plant deep in the soil, grows to the surface, and then develops into the above-ground stalk you see bearing the cap. The underground part when well developed is termed a pseudorhiza (false root). The plant which produces the mushroom is simply a mass of threads (mycelium, pl. mycelia). Hence the fruit body has this same basic structure as the vegetative stage and the spores when they germinate produce a thread which branches to form many, all of which are connected to each other. The tips of the threads are where the chemical activity of liberating enzymes to digest food material takes place.

In the simple fruit body outlined above, the young spores are initiated on cells exposed to the atmosphere and the drying effects of rapid changes in relative humidity. Excessive loss of moisture is the most destructive force with which the mushroom plant must cope in its process of reproduction. Hence it is not surprising to find many species that have developed ways of protecting the young spores. Layers of tissue (veils) have developed which supposedly act to keep the humidity high in the area where the spores are developing. One type of veil, found in *Amanita,* covers the entire young fruit body and may be referred to as the outer veil. If this layer is very tough when the cap breaks through it, the remains are left around the base of the stalk in the form of a sack or cup. This "cup" is termed a volva. Never apply this term to any other structure of the fruiting body.

A partial veil is present on many fruit bodies. It extends from the cap margin to the stalk near the apex of the latter. At the time this veil breaks from further expansion of the cap, the spores are nearly mature. In some species the outer and inner veils become intergrown, in which case the combined remains are often found distributed along the stalk.

One will soon notice in his field observations that the mushroom fruit body is at the mercy of the weather — it dries out rapidly, especially in species with small fruit bodies. In some species, however, the fungus has adjusted to this. The fruit body dries out quickly, but it is tough and as soon as rain or high humidity recurs, it revives and produces more spores. When looking at a fruit body always question the water loss. However, large fleshy fruit bodies are often short-lived, not because of water loss, but because they are food for insects and are soon destroyed by the larvae of these insects. A balance seems to have been worked out here in that many spores are produced before the insects completely ruin the fruit body. This is an item of importance to the person collecting fungi for use as food. The insect larvae commonly enter the fruit

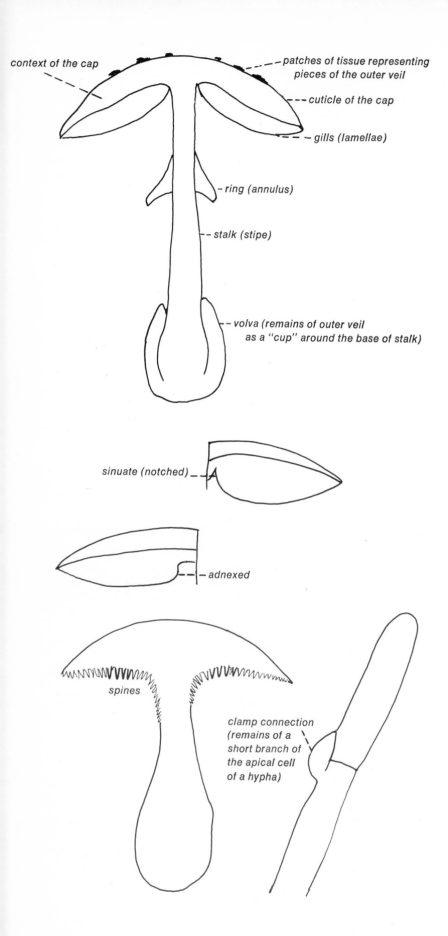

context of the cap

patches of tissue representing pieces of the outer veil

cuticle of the cap

gills (lamellae)

ring (annulus)

stalk (stipe)

volva (remains of outer veil as a "cup" around the base of stalk)

sinuate (notched)

adnexed

spines

clamp connection (remains of a short branch of the apical cell of a hypha)

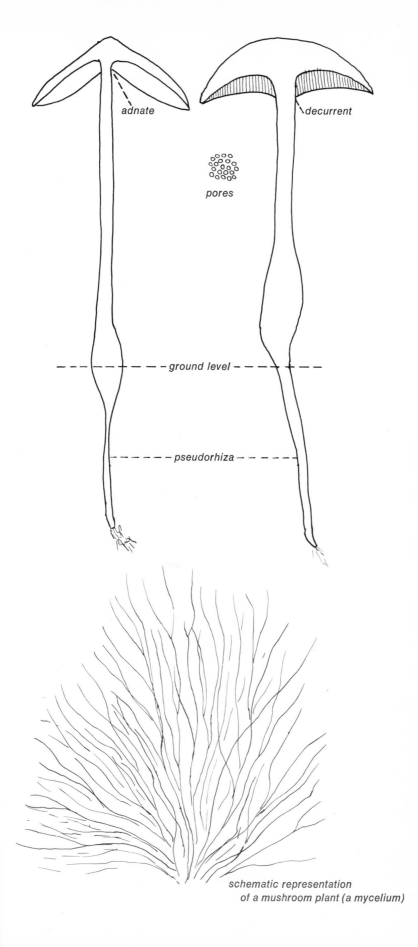

adnate

decurrent

pores

ground level

pseudorhiza

schematic representation
of a mushroom plant (a mycelium)

body in the base of the stalk and work upward into the cap. They work faster when the weather is warm than when it is cool. In other words, if you place whole fruit bodies in a plastic container and carry them around where they will warm up, or if you leave them in your car, the package warms up appreciably, and in an hour the larvae will completely riddle your collections. This is why you should identify your find when you collect it, then cut away any parts showing "pin holes," and above all keep this cleaned collection in as cool a place as possible. If worms have been excluded, you can still get rapid deterioration from the activities of yeast and bacteria, against which there is no ready defense other than to keep the temperature down. This is not as serious a problem in the area west of the Cascades as it is in the Rocky Mountains.

One will quickly note, in the illustrations, that many fruit bodies differ in one or more respects from the true mushroom type (with gills) discussed here. It is on the basis of the features of the hymenophore, along with others, that the different kinds of fleshy fungi are classified. Those Basidiomycetes with gills are in the Agaricales, those with an hymenophore of tubes for the most part are in the Polyporaceae or true pore-fungi. Those with spores produced on spines or "teeth" are the Hedgehog Fungi or Hydnaceae. Many other types will be noted in the illustrations. In other words, the first thing to notice is *the structure of the hymeno-phore.* Be sure to obtain the base of the stalk even though you may cut it off immediately afterwards. Also check carefully for veil remnants (use immature fruit bodies if such are available). Along with these observations note the color or colors of the cap, the hymenophore, the color of the stalk, and any discolorations that develop from handling the fruit body. In addition, check to see whether veil remnants on the stalk leave a ring (annulus) or a sheath (stocking) over the lower half, or whether this sheath breaks up into zones or patches. All these features will contribute to your recognition of the species when you find it again.

The features of the spores are very important in the taxonomy of mushrooms. One feature of importance can be observed without the aid of a microscope; it involves making a spore print. As previously stated, the spores are produced on the spore-mother cells (basidia, sing. basidium) and discharged at maturity. They slowly fall to the ground if the air is perfectly quiet. So, to obtain a mass of spores for identification purposes, cut off the stalk near the apex and place the cap gills *down* on a piece of white paper. Cover the set with a suitable cover to prevent any air currents from reaching the cap and let it stand from one to five hours. This should be time enough to get a good spore fall. Remove the cap and you will observe a white to colored dust on the paper — usually with the outlines of the gills preserved. Allow this deposit to air-dry for 10 to 15 minutes and record the color by comparing it (in daylight) to some well-known colored object or product, such as roasted coffee beans, black chocolate, iron rust, cream, and so forth. But be prepared to find that standardized terms are often used for certain color groups in the literature. Under white-spored, for instance, many species are grouped with pale yellow or pinkish tan spores; under the red-spored group are many in which the deposit was reddish cinnamon; under purple brown are groups of species with red to chocolate black spore deposits; and in the black-spored group red-spored and dull rusty brown-spored species are known. At the present time we are more interested in the exact shade or tint of the deposit for *each species* since the major groupings of species now involve more anatomical characters than were used in older works, such as Kauffman's *Agaricaceae of Michigan.* Spores from the deposit can (should) be used for study under the microscope if this procedure is contemplated.

Since this book is not designed primarily as a technical publication, the reader who wishes to equip himself with a microscope and take up the technical study of the higher fungi is referred to the text of *Mushrooms in their Natural Habitats* by Smith, republished by Hafner Publishing Company of New York; *North American Species of Psathyrella,* vol. 24 of the *Memoirs of the New York Botanical Garden,* Bronx Park, New York; and in particular, *How to Know the Non-Gilled Fleshy Fungi* published by Brown Publishing Company, Dubuque, Iowa.

The Systematic Classification of Plants

In order to refer conveniently to plants each kind must have a name. This consists of the name of the genus (as in *Morchella)* and the species epithet — the word designating the particular kind of plant: *esculenta* for instance, which indicates the kind of *Morchella* under discussion. The name of the species is the combination of the name of the genus (generic name) and the species epithet: *Morchella esculenta.* The language of plant taxonomy is Latin, so the species epithet and the generic name are in Latin. The generic name is a noun, and the species epithet is usually an adjective which agrees with the noun in gender and number, but it may be a noun used in the genetive case, as in *Cantharellus friesii* (the *Cantharellus* of Fries).

A genus may include one species or more than 100. This depends on the opinion of the investigator. Genera are merely groups of species — therefore our two most used groups in the present text are: the *species* and the *genus.* A species may be defined as a population of individuals — which is again a "group."

The group next highest above the genus in our system of classification is the *family.* This consists of a group of genera and in Latin bears the ending *aceae* — Amanitaceae, for instance. Families are grouped into orders, and the Latin ending for an order is *ales* — as in Agaricales. The next highest grouping is the grouping of orders into subdivisions, namely Ascomycotina (Ascomycetes) and Basidiomycotina (Basidiomycetes). The highest category, of course, is the division, which is for our purposes the Fungi. Some people now recognize fungi at the level of a "kingdom" — parallel with the plant kingdom (for green plants).

An outline of the system used here follows:

Fungi — kingdom or division

    Ascomycetes — subdivision

        Discomycetes — order (the ending is an exception to the rule for ordinal endings)

        Hypocreales — order — with the *ales* ending

    Basidiomycetes — subdivision

        Auriculariales — order

        Agaricales — order

            Boletaceae — family

            Amanitaceae — family

                Amanita — genus

                  A. muscaria — species

Note that the puffballs and related fungi are keyed out directly to the genus. This is because so many changes in their classification are still being made.

Techniques

Photography: There are so many collectors in the western area today taking professional quality photographs, that I do not have the temerity to pose as an expert in this field. However, it is considered appropriate to give some data relative to the pictures reproduced here since questions come up repeatedly. My photographs, which, with a few exceptions, are the ones reproduced here, were taken indoors with 3200 or 3400 degree Kelvin photoflood lights, depending on the type of film used. Habitat photographs are difficult to obtain under unfavorable field conditions, and these prevail most of the time in the region if the season is productive of mushrooms. Under such conditions time also becomes a major factor. Dark green or black (velvet) backgrounds are preferred so the color of the background does not overpower the viewer and distract his attention from the fungus.

Solutions: The following solutions are used in making microscopic studies and in testing for color reactions. **Keep all solutions out of the reach of children.**

1. Water (H_2O). Ordinary tap water, or distilled water. Water is the standard mounting medium for fresh material. The appearance of hyphae and spores in water is taken as the standard when comparing the results in other media.

2. Potassium hydroxide (KOH). A 2.5 percent aqueous solution. It need not be exactly 2.5 percent but for fleshy fungi should not be over 3 percent. Authors differ in the concentration used depending on the toughness of the material being studied. For very delicate tissues weaker solutions may be used to avoid too much swelling of hyphal cells (or their actual disintegration). A 2.5 percent solution will give accurate color reactions on fresh material though in the literature it will be noted that stronger solutions are often used for this purpose.

3. Ammonia (NH_4OH). A 14 percent (approximately) aqueous solution is used for color tests both in the field and under the microscope. In Europe the solution is often used for the study of dried specimens.

4. Iron salts — iron sulphate or iron chloride — $FeSO_4$ or $FeCl_2$. A 10 percent aqueous solution is used for color reactions on fresh material but it will also give color reactions on well-dried specimens, usually more slowly however.

5. Melzer's reagent. KI (potassium iodide), 1.5 grams; iodine crystals, .5 grams; water, 22 grams; chloral hydrate, 22 grams. The solution keeps well so it can be made up in larger quantities than indicated above, ten times as much is a convenient amount. **Be sure to keep this solution out of the reach of children.**

Melzer's, as it is commonly called, is used for studying spore walls, hyphal walls, and to a lesser extent cell content. A gray, blue, violet, or purple reaction of the wall or cell content is termed amyloid. This reaction can also be seen on spore deposits on a glass slide simply by adding a drop of Melzer's to the deposit and in a minute or two viewing the treated deposit against white paper. If amyloid, the color of the spore deposit will be blue to violaceous. If the reaction is bay brown (pale or dark) it is termed dextrinoid. If no change takes place or the wall becomes merely yellowish, the reaction is termed nonamyloid (or inamyloid).

Collecting techniques: The first task is to collect specimens for purposes of identification. For all but very minute specimens, I recommend a large market basket with a strong handle and without a cover. Waxed paper (not the very heavy type) is used for wrapping the specimens. Lay the specimen on a sheet of wax paper of appropriate size, roll it up and twist the ends to form a package. These packages can be stood upright in your basket and if carefully handled the specimens will be in perfect shape — even for photography — when you reach home. The packages "breathe" through the twisted ends and so the specimens do not sweat. Always place the large heavy collections in the bottom of the basket. I usually carry an empty four-quart container in the basket for the very small collections.

When collecting for the table (and you can collect for both identification and the table at the same time) carry about 4 large paper bags in the basket, and when you find a cluster of a mushroom, *Pholiota squarrosoides* for instance, detach it, look at it carefully to be sure it is not some other species, and then cut the stalks off at the juncture with the cap (or near it), and examine for "pin holes" (larval tunnels). If there are no pin holes and no soft spots on the gills or cap, then clean off any dirt and put the cap in the sack you have reserved for that species. **NEVER** use plastic bags as containers. I realize that plastic bags are used a great deal, but I disapprove of them because the specimens sweat so readily, especially in hot weather. The two important goals to keep in mind are: When collecting for purposes of identification, do everything necessary to get the specimens home in exactly the condition they were in when you found them still showing all the markings they had originally. When collecting for the table be sure of your identification, and collect cleanly. If the fruit bodies are clean when they are collected they do not need to be washed before use.

Edibility

Always remember that our knowledge of which species are edible has been obtained by the trial-and-error method. For many species the edibility is based on many reports; for others only a single report may be available. This means there is a built in factor of possible error as far as many species are concerned. This is enhanced by the fact that any one person may get a severe reaction from a given species. This is why I advise people trying a species for the first time to eat only a small amount — such as a tablespoon full of cooked mushroom. If no ill effects are noted by the next day, try more — if you like the flavor. There are about 100 edible kinds included in this work and they vary from nonpoisonous to the choice kinds such as *Sparassis.* Every hunter should go about building up his own fund of information in a systematic way. In time one will learn that, in a single species, races or strains are encountered, some of which are more desirable than others. Mushrooms are no different than other plants in this respect. Keep records on exact localities. A mycelium will live for years and produce annual crops. Anyone with known allergies to fungi should keep a careful record of the species he eats and his reactions to them.

As for recipes for cooking mushrooms, I am at a handicap because of my allergies to most species. But for people in the Northwest, there are good booklets of recipes for cooking wild mushrooms based on experience in the area. The Puget Sound Mycological Society, The Oregon Mycological Society, and the San Francisco Mycological Society have all published booklets containing recipes, many of them original. Also, *The Savory Wild Mushroom* by McKenny and Stuntz contains a recipe chapter entitled "Selected Wild Mushroom Recipes." Many of the contributors to these publications are true experts in their area.

A few words of caution: Do not eat raw wild mushrooms promiscuously even though you know the species is edible when cooked. Cases of poisoning have been reported for edible species when eaten raw. However, our commercial species, *Agaricus bisporus,* is frequently used raw in salads. I have never heard of more than an occasional reaction — no worse than my own allergic reaction to the species.

The entire western area contains so many species not known from other regions that one must be very careful in regard to edibility. We have, for instance the case of *Pholiota aurea* as reported by Wells and Kempton from Alaska. It is a good edible species in Europe, but in Alaska a strain occurs that has produced a number of cases of mild poisoning. In Idaho, *P. hiemalis,* a close relative of *P. aurivella,* has been found to be poisonous. My reason for including so many species here for which edibility is not "officially" known, is to help in their recognition, and hopefully set the stage for gaining more exact data on their qualities as food.

In some species, such as *Boletus edulis* the stalk is as good to eat as the cap. In other fungi the stalk is tough and should be discarded. In still others as

Laetiporus sulphureus, cut off the tender growing edges of the fruit body and discard the more fibrous parts. In morels, I prefer to have the stalks cut off but some people use the stalks. One can usually determine from the fresh consistency of the stalk whether it will cook up properly or not.

Drying mushrooms for winter use is a popular method of preservation. In fact, in Switzerland they have developed excellent electric home driers equipped with a slow-speed fan. The secret to proper drying is to have warm (not hot) air moving through the mushrooms laid out on a screen (do not pile them too deep). Several screens supported in a framework and with a hot plate as a source of heat will make a good drier — but operate the hot plate on low heat and have the screens low enough to make proper use of the warm air. I wrap a fire-proofed canvas around the drier used for my specimens. The best results are obtained if the fruit bodies are cut in half lengthwise before being placed on the drier.

Mushroom Poisoning

So much has been written about mushroom poisoning that many people think of mushrooms as being worse in this respect than other groups of plants, the legumes for instance. In the family to which peas and beans belong, there are many species in which the fruits or the leaves of the plant are violently poisonous — but most people do not realize this since their contact with the family is through the edible species important in agriculture. In mushrooms (fleshy fungi as a group) a number of different chemical compounds occur which have different effects on the human body. These have been studied over the years, particularly in Europe, but in recent times much of the basic work on the chemistry of the fruit body has been done at the University of Wasnington at Seattle. It is a matter of interest to note that in the last twenty-five years the best publications on the use of mushrooms for food, and on their basic chemistry have come from our western area, notably the Pacific Northwest. Varro E. Tyler has summarized the present situation relative to mushroom poisoning in a most concise and readable manner in *The Savory Wild Mushroom,* pages 205—12. I recommend this account to all who expect to, or do, eat wild mushrooms.

I have already commented on eating raw mushrooms. Not every authority agrees with me on this point, but I still insist that it is better to be safe than sorry — and I *do not* recommend eating raw wild mushrooms in a region where there are so many species on which our data on edibility are not well documented. I know, for instance, that in Seattle for years *Paxillus involutus* has been collected in Seward Park and eaten — cooked, pickled, and raw in spite of its questionable reputation as an edible fungus. The beginner can well afford to be overly careful.

In the text, and in previous publications, I have used the term "observe the usual precautions": These are: (1) eat only one kind at a time; (2) eat only fruit bodies that were in good condition (free from larvae and lacking soft spots or discolorations caused by bacteria or yeasts); (3) cook the material well; (4) eat only small amounts the first time; (5) never overindulge — such as by eating a quart of cooked mushrooms at one sitting; (6) have each member of the family test each new kind for his or her reactions to it. **But above all be critical in making your identifications in the first place.**

Starting from the most poisonous of the species common in the area and progressing toward the edible, we find the genus *Amanita* at the head of the list. *A. phalloides* has now entered the picture in a serious way. *A. virosa* is likely to do the same any year in the areas where birch is present. Every collector is urged to memorize the features of *A. phalloides.* These are: (1) the sacklike to cuplike volva; (2) a ring on the stalk; (3) white spore deposit; (4) gills free from the stalk at apex; and (4) olive to yellowish cap color. *A. pantherina* is next on the list of poisonous species in the Pacific Northwest, and *A. muscaria* might rate third, but it is seldom fatal, and people who know how to boil the poison out of it, use it as an esculent. Be that as it may, my recommendation is not to eat any *Amanita.* True, some of the species are edible, but there are enough other edible species in the area to more than make up for the loss of a few.

In my estimation a whole group of brown-spored fungi are next on the poisonous species list: *Inocybe* represented by hundreds of species in the West, and the *Galerina autumnalis* group with special emphasis on *G. venenata,* an endemic species. The latter produces the same type of poisoning as the deadly amanitas. Next would follow a miscellaneous number of species from various genera causing "mild" cases of poisoning, mostly of the gastrointestinal type — sometimes in one person but not in another. These cause considerable discomfort to the patient but are seldom really dangerous. Species of *Entoloma,* red-spored agarics with angular spores, would belong here. Indigestible coarse *Lactarii* would come next, and these intergrade with the low-grade edible species.

The problem of allergies is a separate one and must be explored on an individual basis since reactions can be caused by edible mushrooms. From the above, one can readily see that it is extremely important to know which species you are eating, and to eat only one kind at a time. The importance of keeping a card index of each species you eat, and your opinion of them along with any reactions experienced, is self-evident. Also, it should be clear that no one can predict for certain just how any one species will affect a given person. For this reason and because the collector must learn to accept responsibility for his own decisions, neither the author of this book nor the University of Michigan Press accept responsibility for any adverse affects any user of this book experiences as a result of eating mushrooms or other fleshy fungi.

One of the important discoveries of recent years is the poisonous compound in *Gyromitra esculenta*. Simmons (1971) reported on this in his article on mushroom toxins. Monomethylhydrazine (MMH) is the culprit. There is sizable literature on its effects because the compound is used for rocket fuel by the United States Air Force. Since the compound boils at 87.5 degrees C, parboiling the specimens once or twice in a well-ventilated room *usually* removes the poison. With this information at hand, one is strongly inclined to recommend parboiling material of any *Gyromitra* to be used for human consumption, and one might as well include the black morels also.

Thioctic acid has received some attention recently as a treatment for *Amanita phalloides* poisonings but to my knowledge has not been officially approved as a treatment by the FDA. However, poison centers in the area should be alerted to have it in stock for experimental purposes.

Comments on the Flora

If one proceeds from winter conditions through spring to summer and back to winter again, one of the interesting features of our western area is the group of species fruiting in the wake of melting snowbanks or actually at their edges. This flora, if it may be called that, has come to be known as the snowbank flora. It is a fairly sizable flora in the Rocky Mountains of the Salmon River drainage in Idaho, and in the Cascade Mountains. In the southern Rocky Mountains, especially in the San Juan group, the species are often different, as one would expect. It is in this area that I have found morels in late August in areas very recently freed from snow cover. I am not prepared at present to comment on the extent of this flora in the southern

Rockies since I have not followed it from the lower elevations upward. The species comprising the northern Rocky Mountain snowbank flora, however, are in *Lyophyllum, Hygrophorus, Mycena, Cortinarius,* and *Hebeloma* chiefly. Obviously, these genera are not particularly closely related. It seems obvious to me that the species in this group are well established throughout the forest zone, and have adjusted to this fruiting pattern, possibly as a response to the habitat drying out and warming up as summer progresses.

Species of *Rhizopogon,* at least in the northern Rocky Mountains, have two fruiting seasons. The first one comes as the duff dries out from the spring moisture and a few showers wet the surface. In the Priest Lake area of Idaho this fruiting may peak for a ± 10 day period in July, but here microhabitats become all important. Those with ample moisture have the species fruiting later. This aspect is a particularly important one in the Salmon River area of Idaho. Here one may hunt for *Rhizopogon* species from shortly after the fourth of July to past the middle of August. A second heavy fruiting comes late in the fall after heavy rains. Light frosts do not affect it. Many species fruit during both seasons.

In the central and southern Rocky Mountains the heavy fruiting of fleshy fungi generally comes during the periods of high precipitation in July, August, or early September, and one must hunt to find the right "spot" if he is to be successful. It may rain in one mountain range and those on either side of it be dry. I shall never forget the summer of 1950 when we located at the University of Wyoming Science Camp in the Medicine Bow Mountains and then made three trips a week for much of the period over to the Pole Mountain Range east of Laramie to collect mushrooms because that was where it was raining. The fall season in the central and southern Rocky Mountains is cut short by cold weather because of the high elevations.

However, in the Priest Lake area of Idaho south to the Boise River, the months of September and October at times witness major fruitings of fleshy fungi — always with variations depending on the weather.

The area from the crest of the Cascades to the coast makes up its own rules for the behavior of fleshy fungi as the season progresses. One year fruiting on the coast may be early — in part at least stimulated by fog. In another year the major fruiting may be late in November, or at times there seems to be no true periodicity, fruiting occurs in a desultory manner. The western slope of the Cascades is often best in October. In southwestern Oregon, the best month is usually November. In both areas if the fall rains come early

(September) there will be a major fruiting of *Russula, Suillus* (especially *S. lakei),* and species of *Mycena.* Usually about two weeks later the *Cortinarii* come out in all their colors and dominate the forest habitats. When these have all but disappeared there may come a luxuriant fruiting of *Tricholoma* species especially under lodgepole pine. Wood-inhabiting species start fruiting shortly after the fall rains start and continue until winter, but not always in abundance.

According to my experience, the makeup of the fruiting follows a pattern in this way: If, let us say, *Naematoloma fasciculare* starts to fruit, it often continues to do so through the entire season. The other common species of *Naematoloma* in the area, *N. capnoides,* may be rare that year. During some other season the proportions will be the reverse. Certain seasons give the impression that mushrooms are everywhere, but when you study the fruiting you find that it is made up of relatively few species. Another season may furnish one with almost untold diversity, as happened in 1947 when I collected 192 species of *Cortinarius* in the Mount Hood National Forest. A feature of second growth Douglas fir stands is the abundance of *Inocybe* species, as was observed at the D. E. Stuntz Foray in 1972 in the Cispus Environmental Center south of Randle, Washington. The heavy fruitings of *Cortinarius* are found in relatively open habitats among mosses at elevations of 1500 feet to timberline. The coastal jungle of salmon berry, salal, devil's club, vine-maple, and alder, which has, for instance, taken over most of the cutover land in the Siuslaw National Forest, features a miscellaneous assortment of species very difficult to find because of poor visibility to the ground level. Even the wood-dwelling species in this area often seem to fruit poorly, the fruit bodies being small and occurring in small numbers. In this area, however, there is often an abundance of very small mushrooms — with caps 2 to 12 mm broad and threadlike stalks. When the undergrowth in such areas is cleared out, in a year or so the larger fungi usually fruit in abundance. In other words they have been living there all the time but the stimulus to produce large fruitings appears to have been lacking. This plant association, if one may use the term for the coastal jungle, however, produces the largest number of species of *Lepiota* of any habitat in the region. The best areas for *Galerina,* on the other hand, are near and above timberline on moss or mossy logs. Also the best collecting for this genus has been in dry relatively warm years. When collecting this genus, one soon learns the meaning of the term "microhabitat."

In summary, one might say that the variations in

weather from season to season superimposed on the various habitats at different elevations and exposure create a situation for the mushroom hunter very similar indeed to that which makes fly fishing such a fascinating sport: skill and good luck must combine to place a collector in the famous 10 percent of the hunters who have 90 percent of the success.

One of the outstanding features of the flora in the entire western region is the evolution of species of fungi in relation to the species of conifers (the dominant forest type) occupying the area. In one genus, for instance, namely *Rhizopogon* (see Smith and Zeller, 1966), the number of species found is over five times that known for any other area in the world regardless of its size, and the study is not yet complete. As I often explain to visitors, I still have the world's largest number of unidentified *Rhizopogon* collections.

Almost all the species described from the western area appear to form mycorrhiza with conifers, and some occur rather consistently with a single species, *R. parksii* with Sitka spruce, *R. ochraceorubens* with lodgepole pine. Many of these fungi can be collected in quantity if one collects in the right place during the proper season of the year, and some may turn out to be highly desirable edible species. The few I have tried were good. Why do we have so many more species of *Rhizopogon* in our western flora than occur in any other part of the world? The problems which *Rhizopogon* poses for fungous geography, species differentiation, ecology and physiology, are, literally, too numerous to discuss here. However, it is pertinent to ask if other groups of fungi follow the same pattern. The answer is definitely yes, but a variety of patterns occur. *Cortinarius* is an example of a genus with hundreds of species, many endemic, which grow in association with western conifers. But here we are dealing with a group of over 800 species for North America with a large representation of species in hardwood forests as well. This situation is not exactly parallel to that of *Rhizopogon,* since *Cortinarius* occurs abundantly in both conifer and hardwood forests. The family Gomphidiaceae, however, parallels the *Rhizopogon* pattern exactly. Its species practically all form mycorrhiza with conifers, and more species are found in our western area than anywhere else, in fact, they form a conspicuous element in our western flora.

A third genus associated with conifers and exhibiting great diversity in our western flora is *Suillus,* a group of boletes. Pine forests, in general, support a luxuriant *Suillus* flora, with different species associated with different species of pine in many instances, such as *S. sibiricus* and *S. borealis* with *Pinus monti-*

cola, and *S. pungens* with Monterey pine. Western larch has an assortment of fleshy fungi living and forming mycorrhiza with it. *Suillus cavipes, Hygrophorus hypothejus, Gomphidius maculatus, Hygrophorus speciosus, Fuscoboletinus aeruginascens,* and *Suillus grevellei* are a few. Hemlock, the true firs, and Douglas fir *(Tsuga, Abies,* and *Pseudotsuga)* all have large numbers of species of various genera forming mycorrhiza with them. This close relationship of a species of fungus with a single species (or genus) of tree is very important to the mushroom hunter because it is a major cue for him to find his quarry. But he should always remember that there are just as many species which seem to be very generalized in their tree associates. *Amanita muscaria,* including its varieties and races, forms mycorrhiza with spruce, pine, aspen, fir, and, on occasion, with oak.

The oak-manzanita stands in southern Oregon and California offer quite a different type of habitat than do the conifer forests, and one finds there a rather strikingly different fungous flora. *Amanita calyptroderma* for instance, appears in fair abundance — enough to make it important as an edible species. *Boletus satanus, B. regius,* and *B. flaviporus* are other species occurring in this vegetational type. But the detailed mapping of the distribution of species of fleshy fungi in California, and for that matter obtaining a proper inventory of the kinds growing there, is, on a modern basis, still in the beginning stages due to the neglect of this group of organisms by institutions of higher learning in that state during the first sixty years of this century.

Other relatively nonproductive vegetational types occupying large areas of western land are the juniper areas and those covered by sagebrush. Both these areas have interesting floras of fungi, but it is quite another matter to collect species in them at the right time. A feature of the juniper forests, if they can be so designated, is the abundance of species of *Geastrum, Tulostoma,* and species of puffballs in the order Lycoperdales. Sagebrush areas, at the right time during a good season, also feature Lycoperdales, but in addition a number of species of *Agaricus.*

The size of our western flora of higher fungi is not yet known. I hope to have around 4,000 species in my proposed western manual when it is finished. But I feel certain that many endemic species will still remain "undiscovered" by mycologists. This does not mean that "no one" knows about them, as I learned from personal experience when collecting fungi which fruit underground in the Sawtooth Mountains of Idaho. One rancher invited us to hunt on his ranch, but before going out he suggested we look in his horse corral. We did not show enough enthusiasm (apparently) so he took us there, and we found a species of *Macowanites* growing in abundance. It was a "new" species in the sense that it was undescribed, but this rancher had been eating it for as long as he had owned the ranch and could tell us about its edibility, seasonal occurrence, how the top of the cap split open in a characteristic pattern, and that the cap color was extremely variable.

If one simply considers all the major vascular plant associations of the area occurring through the various life zones, the erratic weather patterns (which determine the fruiting of these fungi) along with the handful of qualified mycologists in the country capable of doing a critical study on them, the reason why a book with "all" the species in it has not been written should be obvious to all. In spite of the intensive work I have done in the western area, I consider this as just a good beginning. In view of the magnitude of the task it seems only reasonable to "chip away" at it as best one can. In terms of publications, it should be abundantly clear that a work on the regional flora is badly needed.

One question that is frequently asked is: Why are there so many endemic species of fungi in our western area? My answer is that the same forces that have created such a diverse flora of seed plants have also brought about the diversity in the fungi. The size of the area must always be kept in mind — there is room for plants to move around and to become isolated to some extent, and this in time produces varying degrees of diversity in the vegetation. The age of the area is of great importance, because the longer the time span, the more time there is for isolated populations to develop features different from the original stock. The orientation of the major mountain ranges is also thought to have had an influence on the vegetation throughout geologic time.

It is well known that there have been great climatic changes in the western area. During cold periods as the glaciers built up in mass and moved southward, they moved down the mountain valleys, not across them. The ranges are oriented mostly in a north-south direction, and the vegetation (or segments of it) at any one period was capable of surviving to some extent by migrating slowly southward. As the ice melted, these remains of the earlier vegetation were able to reseed the areas vacated by the ice and hence the vegetation gradually moved northward. From the standpoint of speciation (the production of a flora rich in the number of species) these gradual movements meant that at no time was all the vegetation wiped out, as presumably happened in Europe as the ice moved down to the Alps. In our western area as the ice melted, there must have been rapid evolution of new species in the flora which then competed with those species remaining of the original flora. I am quite convinced that such a situation as this accounts for the great diversity of species of fleshy fungi in our western area when compared with the rest of the land masses of the earth.

How to Use the Keys

A key in a taxonomic work is a device to enable a person unfamiliar with a particular group of organisms to eliminate, in an orderly manner, all the species except one. It is purely a mechanical device and not meant to indicate the degree of relationship between species or genera. Attempts to express the author's ideas as to relationships of the species are usually done by the arrangement of the species in the text of his work. Ideas on the relationships of organisms are largely hypothetical and are based on correlation of characters considered to be relatively stable. The weakness of every key which includes only a few species of a genus or other group, *Boletus,* for instance, is that most of the time one will find a species not in the key. So if at the end of the key you come to a choice (a species), the description of which does not check with your fungus, the logical conclusion is that your fungus is not in the book.

To use the key start with choice 1 in the Key to Selected Major Groups of Fungi and read both entries under that number. Then select the entry which best describes your specimen and proceed to the group or number indicated. If, for instance, the bolete shows a netted pattern over the stalk, go to the entry so indicated in the key and repeat the procedure. Continue in this way until you come to a species name. Then compare the description of that fungus with the characters of the collection you have. If there are serious discrepancies, one can conclude that the fungus is not in the book. For instance, the description may call for a cap 5–15 cm broad and your collection has one 5–7 cm broad. This is not a serious difference. But if the cap in your collection is only 1–3 cm broad at maturity, the difference *could be* important. In other words, one must learn to use judgment in evaluating the taxonomic characters of the organisms being studied whether they be mushrooms or some other type of organism. One's judgment develops with experience — it is much easier to see important features after you have examined 200 different species (and each one several times) than when you have seen only a dozen or less.

Key to
Selected Major Groups
of Fungi

1.   Fruit body discoid, cup-shaped with or without a stalk, or in convoluted masses with individual fruit bodies not clearly defined (see Auriculariales also) . . . .
. . . .(p. 27) The cup-shaped *Discomycetes*

1.   Not as above 2

     2.   Fruit bodies capitate-stalked to pileate-stalked; the spore producing cells born on the upper (or outer) surface of the cap or head rather than on the under surface, the head or cap varying from smooth to wrinkled to pitted . . . .(p. 27) The stalked ± capitate *Discomycetes*

     2.   Not as above 3

3.   Fruiting structures (perithecia) born in a layer covering the hymenium or other surfaces of a fruit body of a mushroom or polypore . . . .(p. 45) *Hypocreales*

3.   Not as above 4

     4.   Fruit bodies very tough-cartilaginous, saucerlike to ear-shaped, dull brown; growing on conifer logs with the bark still on them (especially on fir logs in the West) . . . .(p. 49) *Auriculariales*

     4.   Not as above 5

5.   Fruit body tough, woody, semifleshy to truly fleshy; spore-bearing surface in the form of tubes (visible as pores on the underside of the cap), or spines which hang down from the cap or from a lump of tissue termed a tubercle . . . .(p. 50) *Aphyllophorales*

5.   Not as above 6

     6.   Fruit body a ± brittle upright club or cylinder, or profusely branched with flattened to round (in section) branches; apex enlarged (by radial growth) into a cap in *Cantharellus* (which has obtusely edged gills) . . . .
. . . .(p. 63) *Cantharellales*

     6.   Not as above 7

7.   Fruit body fleshy to pliant; spore-bearing surface in the form of gills or tubes, but if tubes are present they usually separate easily from the cap and feel somewhat gelatinous when pressed between the fingers . . . .
. . . .(p. 75) *Agaricales*

7.   No spore bearing surface readily evident, the spores are born in the interior of the fruit body and may or may not be exposed or discharged at maturity . . . . . .
. . . .(p. 245) *Puffballs and Related Fungi*

Discomycetes

The Discomycetes are a large and diverse group of Ascomycetes. Only a few of the conspicuous species are included here.

Key to Genera

1. Fruit body about the size of a silver dollar or these merged into a crust; attached by coarse fibers (rhizoids) to the burned soil on which it grows . . . . (p. 28) *Rhizina*

1. Not as above 2

   2. Fruit bodies stalked and with a cap or head at the apex, or a cup which becomes ± expanded 3

   2. Not as above 6

3. Cap attached only to apex of stalk . . . . (p. 34) *Verpa*

3. Cap intergrown with stalk for part of its radius or all of it 4

   4. Head pitted . . . . (p. 35) *Morchella*

   4. Not as above, head or cap folded or convoluted 5

5. Cap (or cup) pallid to gray or blackish, surface wrinkled to ± smooth . . . . (p. 39) *Helvella*

5. Cap or head ± smooth to very convoluted, color dull yellow to red brown or bay . . . . (p. 41) *Gyromitra*

   6. Fruit body ± saucer-shaped to broadly cup-shaped; spore producing surface usually soon becoming uneven to wrinkled; stalk short, thick, poorly developed; spores with a projection (apiculus) at each end . . . . (p. 30) *Discina*

   6. Not as above 7

7. Fruit body black, large, ± cartilaginous . . . . (p. 32) *Sarcosoma*

7. Not as above 8

   8. Hymenial surface dull lilac at maturity; cups large and thick-walled, splitting into segments as they expand . . . . (p. 31) *Sarcosphaeria*

   8. Not as above 9

9. Fruit body typically cup-shaped but occurring in large clusters with individual cups misshapen from mutual pressure (those on the periphery often remaining saucerlike) . . . . (p. 28) *Peziza proteana* f. *sparassoides**

9. Fruit body more or less urn-shaped; exterior pale brown; hymenial surface pale pink when young . . . . (p. 33) *Neournula***

*At this point in the key it is impractical for the purposes of this work to define the genus *Peziza*.
***Neournula* contains only the species illustrated.

Rhizina

The treatment of this single species is sufficient for both a genus and a species diagnosis.

1 Rhizina undulata

Field identification marks. (1) It grows on recently burned areas (especially where forest fires occurred the year before); (2) the fruit body is rather shapeless and seems to form a crust over the charred earth; (3) the presence of numerous "rootlike" structures (rhizoids) anchoring the crust to the substrate.

Observations. This is a most peculiar Ascomycete with most peculiar habits. It is said to be parasitic on pine seedlings in England. In North America it follows forest fires but will also appear where brush piles have been burned. It is widespread throughout the Northern Hemisphere in the spring and fall.

Edibility. Not recommended.

When and where to find it. See Observations.

Microscopic characters. **Asci** 350–400 x 16–22 μ, operculate. **Spores** 25–40 x 8–12 μ, fusiform, apiculate at the ends. **Paraphyses** narrowly clavate, the apex often thinly incrusted.

Peziza

Most large cup-fungi were placed in *Peziza* at one time or another, and, in the currently restricted sense, most of the large nongelatinous species with nonreticulate spores are still placed there. The one illustrated does not answer this concept well because the cups are mostly modified in shape because of mutual pressure.

2 Peziza proteana f. sparassoides

Field identification marks. (1) The aspect of the cluster as one sees it growing immediately recalls the genus *Sparassis*. (2) The color varies from lilac around the base of many of the shallow cups when they are young and fresh to pinkish tan on aging, and finally dingy pallid as more pigment is leached out. The shallowly cup-shaped fruit bodies occur around the edge of the cluster where mutual crowding has not caused distortion.

Edibility. Edible and rated highly.

When and where to find it. It is probably more common than the records indicate. It is considered rare in Oregon, and the same appears to be true in other areas, but it is, nevertheless, widely distributed over the land masses of the Northern Hemisphere. It fruits during the summer and fall.

Microscopic characters. **Asci** 8-spored, cylindric, operculate, 182–230 x 7.8–10 μ; wall blue in Melzer's especially around the operculum. **Paraphyses** septate, 3–4 μ wide at the base, 4–7 μ at the apex. **Outer layer** of apothecium composed of more or less globose cells with some interwoven hyphae; **middle layer** of interwoven hyphae, and the interior region of inflated hyphal cells, the latter merging into the hymenium.

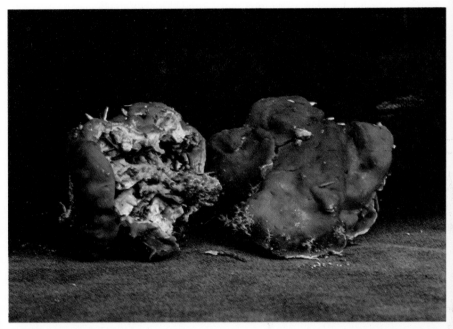

1 *Rhizina undulata*   About natural size

2 *Peziza proteana* f. *sparassoides*   About one-half natural size

3 Discina perlata

Field identification marks.   (1) The pallid exterior and the often stalklike base; (2) the saucer-shaped fruit bodies with their convoluted hymenial surface; (3) the dull to dark cinnamon colors. The cups develop rapidly but often mature their spores slowly so it is common for collectors to pick up "sterile" specimens. It grows under conifers, often around old stumps.

Observations.   The species illustrated here is offered as an example of the genus. Any one interested in identifying specimens to species should consult the paper by McKnight (1969) which deals in a large measure with material from the western United States.

Edibility.   Not recommended.

When and where to find it.   *D. perlata* fruits early in the spring and may be found as a member of the "snowbank" flora in our western mountains. It is often accompanied by *Gyromitra esculenta,* and is, indeed, related to that genus.

Microscopic characters.   **Spores** 25–35 x 8–16 μ, apiculus 1–3.5 μ long(there is one at each end), smooth. **Asci** cylindric, 380–450 x 18–21 μ; lacking a positive iodine test, operculate. **Paraphyses** cylindric, clavate above, branched below, 225–250 x 6–9 μ (above the branches), 4–5 μ below them, septate, bright yellow in KOH. **Hymenium** 350–450 μ thick. **Medullary excipulum** compact, hyphae interwoven, containing scattered oleiferous hyphae for the first 350–450 μ below the hymenium. **Ectal excipulum** 100–150 μ thick, two-layered.

3 *Discina perlata*   *About natural size*

Sarcosphaeria crassa   4

Field identification marks.   (1) Whitish to grayish cups (the term *volva* should never be applied to them) the size of a tennis ball or larger; (2) their dull lilac to lilac brown interior when mature or nearly mature; (3) habit of originating in the ground and breaking through but they may be superficial on hard, packed soil; (4) the manner in which the cup splits into lobes from the apex downward.

Observations.   In most books the species is found under *S. eximia* or *S. coronaria.* This species is one of our largest western *Discomycetes.*

Edibility.   Not recommended. Some people are poisoned by it; at its best it "leaves something to be desired" as an edible species (see Stuntz, *The Savory Wild Mushroom,* p. 195).

When and where to find it.   It is a common and abundant fungus in early summer in the Salmon River drainage of Idaho, but it is widely distributed in the West.

Microscopic characters.   **Asci** 8-spored. **Spores** 14–18 x 7–9 μ, with two oil drops. **Paraphyses** 5–8 μ wide at apex, branched, septate, constricted at the septa.

4 *Sarcosphaeria crassa*   *About one-half natural size*

5 Sarcosoma mexicana

Field identification marks. (1) The large size of the fruit body (6 cm or more broad); (2) the dark sooty brown to black exterior; (3) the rubbery-cartilaginous consistency; (4) the smoky brown wavy hymenium; (5) the ± stalked base.

Observations. There is a second large black species, *S. globosa* which is readily distinguished from the above by having a much more watery fruit body, the water being held mostly in a broad jellylike base. It is not cartilaginous as is *S. mexicana* — which never "leaks" water as one collects it or carries it in a collecting basket.

Edibility. Not recommended.

When and where to find it. For *S. mexicana:* On old sticks in wet places, late summer and fall in the coastal region. For *S. globosa:* On humus with only the hymenium above the duff at times, under spruce, fir, or pine; not uncommon in the spring in the mountains of the Salmon River area of Idaho.

Microscopic characters. **Hymenium** reddish brown to smoky brown, usually lumpy or otherwise very uneven; in age or when dried cracking to expose the whitish substance of the hypothecium. **Asci** 8-spored, 250–300 x 12–18 μ. **Paraphyses** filiform slightly thickened above and brownish. **Spores** 25–34 x 8–10 μ, hyaline.

5 *Sarcosoma mexicana*   *About natural size*

Neournula pouchetti 6

Field identification marks. (1) The pale brown exterior; (2) the shape resembling a small urn; (3) the lining of the interior of the cup pale pink at first; (4) the short whitish stalk.

Observations. The warty ascospores and the lack of a gelatinized layer in the fruit body are diagnostic microscopic characters.

Edibility. Not tested, as far as I am aware, and not likely to prove worth eating.

When and where to find it. Scattered on humus under various conifers in the Pacific Northwest, summer and fall. The stalk and part of the urn are often sunken in substrate.

Microscopic characters. **Asci** 8-spored, suboperculate, 290–390 x 12–15 μ, bases blunt, usually twisted, occasionally lobed, each connected to the hypothecium by a relatively narrow hypha; the tips often bent; not maturing simultaneously; opening eccentrically. **Ascospores** 25–31 x 8–10.5 μ, ellipsoid to oblong, becoming ornamented with low warts readily stained in cotton blue. **Paraphyses** narrow, septate, freely branched and anastomosing, 3–3.5 μ broad, the tips embedded in a pale brown amorphous material. **Medullary excipulum** about 500 μ thick, of interwoven hyphae 2–5 μ broad. **Ectal excipulum** pseudoparenchymatous.

6 *Neournula pouchetti*   *About natural size*

7   Verpa bohemica
(Early Morel)

Field identification marks.   (1) The cap is attached at the apex of the stalk; (2) its surface is folded lengthwise into ridges and valleys often irregular in outline, or at times the valleys connected by cross veins to form long narrow pits; (3) the pale to dark yellow brown color of the cap; (4) the early appearance of the fruit bodies — usually just before the trees leaf out in the spring.

Observations.   Two forms occur, a giant form and a small (typical) form but the number of intermediates which occur strongly suggest that the cause of the difference is nutritional rather than genetic. A smaller species with a smooth cap is *V. conica.*

Edibility.   Edible for some people but not for others, and do not eat generous helpings on successive days. Also, parboil as a preliminary to final preparation for the table (see the account of poisoning under *Gyromitra esculenta).* One of the first symptoms to appear in *Verpa* poisoning is lack of muscular coordination.

When and where to find it.   In the West it is found in the stream valleys and on low land where alder-cottonwood stands occur. Here it may appear from late winter through early spring, a rather prolonged season.

Microscopic characters.   **Asci** 2-spored. **Spores** 60–75(80) x 15–20 μ. **Paraphyses** clavate, 7–8 μ wide near apex, usually with a brown content in fresh material.

7 *Verpa bohemica*   *About natural size*

Morchella is identified by the pitted head which in all species except *M. semilibera* appears to be a continuation of the stalk (see a longitudinal section). In *M. semilibera* the head appears to be attached part way down the stalk. My own observations on very young fruit bodies show that the head is at first attached at the apex of the stalk, but because of rapid expansion (growth) of the fertile tissue a hollow area forms at the apex of the stalk as the cap elongates upward and at the margins downward. At maturity the cap *appears* to be attached to the stalk ± midway between its lower edge and the apex. The stalk is hollow and this hollow soon becomes continuous with the hollow mentioned above. If this is true, *M. semilibera* should be placed in *Verpa*. I can see no justification for a third genus, one intermediate between *Morchella* and *Verpa*.

Key to Species

1. Cap margin free from stalk for a distance of over 5 mm; cap strongly longitudinally folded or with elongate pits
    . . . . (p. 35) *M. semilibera*
1. Not as above (cap and stalk ± fused) 2
    2. Pits of cap irregular in shape, mostly as long as wide; ridges not blackening . . . . (p. 36) *M. esculenta*
    2. Pits of head ± elongate; the ridges blackening or entire head black from early youth on . . . .
        . . . . (p. 37) *M. angusticeps* and *M. conica*

Morchella semilibera   8
(Half-Free Morel)

Field identification marks.   (1) The cap is free from the stalk for about the lower half of its length from the lower edge to its apex; (2) the cap surface features elongated pits (ridges and valleys); (3) cap color pale to medium yellow brown but in aging the ridges become blackish brown; (4) the shape and size of the head change markedly from youth to age (in age the head is more conic bell-shaped and smaller); (5) stalk finally 6–10 cm or more long, equal at first but in age often flared markedly at the base.

Observations.   Some recent authors use the name *Mitrophora semilibera,* or *Mitrophora hybrida* for this species.

Edibility.   Edible, but not one of the highly desirable species of morel.

When and where to find it. In the West it fruits early in the spring in the cottonwood-alder stands, as for *Verpa bohemica,* but *usually* appears as the *Verpa* finishes its fruiting period. The time varies with the season — as for most fleshy fungi. **Microscopic characters.** **Asci** 8-spored. **Spores** 20–25 x 10–15 μ. **Paraphyses** clavate.

8 *Morchella semilibera*   *About natural size*

---

9   Morchella esculenta
(Common Morel or Sponge Mushroom)

Field identification marks.   (1) The head (pitted portion) is not separate from the stalk, but rather appears to be a continuation of it; (2) the head features a pitted surface with pits irregular in shape and in young heads often rather deep (they become shallower in age); (3) the ridges, typically, are paler than the valleys.

Observations.   In the genus *Morchella* the cap and stalk have fused for part or nearly the entire length of the fertile part, which under this condition is more properly described as a head. There is great variation in the common morel and one can find in the recently published literature some strikingly different morels under the name *M. esculenta.* The concept adhered to in the present work is the one most commonly used in the North American literature. *M. crassipes* is difficult to distinguish from *M. esculenta* when the latter is young. In age it is much larger and the head more elongated. It is as good or better for the table than *M. esculenta.*

Morchella esculenta   *About two-thirds natural size*

Edibility.   EDIBLE and CHOICE. Numerous attempts to produce fruit bodies under controlled conditions commercially
have failed.

When and where to find it.   It fruits in the spring in a variety
of habitats (old orchards, railroad right-of-ways, lawns, beech-
maple forests, oak forests, swampy ground under jewelweed,
beside trails). May is "Morel Month" in Michigan but in the
West they may fruit earlier or later, depending on the precipitation, elevation, and latitude.

Microscopic characters.   **Asci** 8-spored. **Spores** 20–25 x
11–15 μ, ellipsoid, yellowish in deposits. **Paraphyses** enlarged near apex up to 15 μ, septate.

Morchella angusticeps and   10
Morchella conica
(The Black Morels)

Field identification marks.   We have two extremes of black
morels in North America; a narrow-headed morel which becomes black, and a large-headed species which is black
from very early stages on. The narrow-headed form (or
species) *M. angusticeps* (1) fruits early in the spring; (2) has
a gray to dull tan head with distinctly elongate pits which
blacken on the ridges *before* the fruit body is mature. The
M. conica (large-headed black morel) is (1) broadly conic to
subglobose at first; (2) blackish overall when still very young;

10 *Morchella angusticeps*   *Slightly less than natural siz[e]*

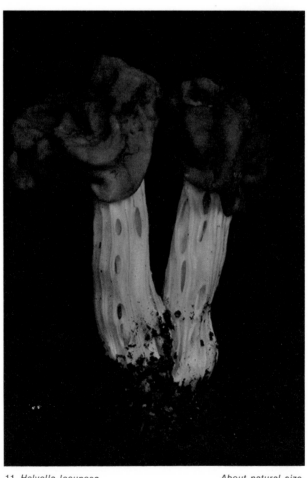

11 *Helvella lacunosa*   *About natural size*

(3) attains the size of *M. esculenta* or *M. crassipes* (up to 18 cm high and 8–10 cm broad).

Observations. There is a need for critical research by local mycologists in the West to determine the "limits" of species in the black morel group. The above classification is strictly preliminary since my field experience in the West at the time of year fruiting occurs is very limited.

Edibility. Questionable! One mycologist in the Northwest has stated to me personally that this group causes more cases of mushroom poisoning in his region than any other fleshy fungus. Also, more people eat them. But there is a problem here that should not be passed over lightly.

When and where to find them. They occur in conifer and hardwood forests during the spring and summer — depending on elevation — and occasionally in the fall — as on the Sundance burn in northern Idaho. Recent burns are the choice gathering places for most collectors because it is here that the black morels can often be collected by the bushel — however *M. esculenta* may occur in burns also. I have found them in August above Trout Lake in Colorado at near timberline. They are found so generally over the area in conifer forests, it seems useless to try to pinpoint habitats other than burns a year or two after the fire.

Microscopic characters. None is included here because from a scientific standpoint I do not believe that valid concepts have been established for the populations in the western area.

Helvella (False Morels)

The characters for distinguishing between *Helvella* and *Gyromitra* are technical and involve ascus and fruit body development. The best field characters are the white, pallid gray to black coloration of the *Helvella* fruit body and the yellow, tan to bay or darker dull red brown of *Gyromitra*.

Key to Species

1. Head ± saddle-shaped; stalk not distinctly discolored at base . . . . (p. 39) *H. lacunosa*
1. Head ± globose-wrinkled; stalk soon dark vinaceous brown over base . . . . (p. 40) *H. californica*

Helvella lacunosa   11

Field identification marks. (1) The elongated stalk with fold-like ridges and with holes (lacunae) in it; (2) gray brown color of the cap (it varies from nearly black to practically white); (3) the typically more or less saddle-shaped cap — variations of this are shown in the photograph.

Observations. The species epithet *lacunosa* refers to the holes in the stalk.

Edibility. Edible and popular with many people; but do *not* eat it raw. I do not recommend it in view of the disturbing situation which has developed with the black morels. One who does elect to try it should parboil the specimens as recommended for *Gyromitra.* Before cooking, throw out any fruit body which shows a white moldy growth. The mold can develop in a few hours if the specimens are in a plastic bag in a warm place (± 70° F).

When and where to find it. Scattered to densely gregarious on soil and humus, late summer on into winter where the weather is above freezing. I have seen it most abundantly on old out-wash areas of mountain streams. It occurs throughout the West.

Microscopic characters. **Spores** (16.5)17.5–21(22) x (10.5) 12–14 μ, ellipsoid, with one oil drop, smooth becoming verrucose-rugulose in age. **Hymenium** 250–380 μ thick. **Paraphyses** hyaline to light brown, pigment when present in the wall, 4–10 μ wide, clavate to subcapitate.

---

12   Helvella californica

Field identification marks. (1) The short stalk consisting of a thin layer of tissue folded into high ridges and broad deep valleys; (2) at the base of the stalk wine red to purplish discolorations soon develop; (3) the "aspect" of the fruit body as shown in the photograph.

Observations. The fruit bodies at times attain a very large size (± 12 inches) and are then very fragile. *Helvella*

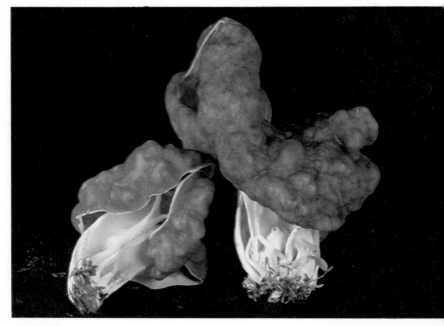

12 *Helvella californica*   *About one-half natural size*

sphaerospora is an eastern species practically identical in appearance with *H. californica* but occurring on rotten hardwood logs. Some authors now place both of these in a third genus, *Pseudorhizina* which name, when translated, means a false *Rhizina*. The name *Pseudorhizina* is a terrible misnomer since these two *Helvella* species have no real resemblance with *Rhizina* (compare the illustrations).

Edibility.   Not much is known, and I do not recommend trying it. *NEVER* eat it raw. If one is tempted to try it, parboil and discard the water before making the final preparation.

When and where to find it.   This is characteristically a western species appearing on soil, near rotten conifer logs, and especially along the scars left by dragging logs out of the forest. It fruits from early spring on into the summer at various elevations. It is common in the Pacific Northwest.

Microscopic characters.   **Asci** 150–200 μ long, 8–10 μ wide, 8-spored. **Spores** 15–18 $\times$ 7–9 μ, smooth. **Paraphyses** 6–12 (14) μ wide at apex.

Gyromitra

See *Helvella* for distinguishing characters.

Key to Species

1.   Stalk in cross section showing interior folds . . . .(p. 41) *G. gigas* and *G. fastigiata*

1.   Stalk not as above (with a single central cavity, or this obscured by the stalk becoming compressed) 2

  2.   On wood or humus, or naked soil such as along roads, fruiting mostly in the fall; head typically saddle-shaped . . . .(p. 43) *G. infula*

  2.   On humus in the spring; head ordinarily not saddle-shaped . . . .(p. 44) *G. esculenta*

Gyromitra gigas and G. fastigiata   13
(Snow Mushrooms)

Field identification marks.   (1) Fruiting in mountain conifer forests usually near melting snow banks; (2) dull yellow to tan head; (3) very short thick stalk with interior folding; (4) general aspect as shown in the photograph; (5) fertile tissue crushed in KOH gives a yellow color.

Observations.   The name *Gyromitra montana* has been given by the Europeans to the American "Snow Mushroom." Note: The photograph is of *G. fastigiata.* For *G. gigas,* see plate 13 in *The Mushroom Hunter's Field Guide.*

13 *Gyromitra fastigiata*   *About one-half natural s.*

Edibility. Edible and choice, but to be safe do not eat it raw. Observe the instructions relative to parboiling as given for the genus. It is one of the very popular species in the Pacific Northwest. *G. fastigiata* has been found to be common in the area. *G. gigas* has oval spores and *G. fastigiata* subfusoid, apiculate spores. They are commonly confused by collectors in the field, but both are edible after parboiling.

When and where to find it. As indicated above, it occurs near melting snowbanks in conifer forests or where snowbanks have just disappeared. It (and *G. fastigiata*) are common species in June and early July in the mountains around McCall, Idaho, and apparently throughout the Northwest.

Microscopic characters. **Asci** 8-spored. **Spores** 26–36 × 14–16 μ, with one large oil drop in addition to smaller ones, ovoid to ellipsoid. **Paraphyses** branched, septate, apex often capitate and up to 9 μ broad, content yellow in KOH.

Field identification marks. (1) The saddle-shaped cap (or some variation of this shape); (2) stalk with a simple hollow (compressed stalks may show 2 small compartments); (3) surface of the head wavy to somewhat irregular but not with brainlike convolutions; (4) the dull reddish brown stalk.

Observations. Both *G. infula* and *G. esculenta* when very young have a cup-shaped "head" (apothecium) but this soon grows out into the shapes shown. Hence at some stages of development there is considerable resemblance between them.

Edibility. POISONOUS, as for *G. esculenta,* never eat it raw. Apparently the poison can be removed by parboiling once or twice and each time discarding the water; but I do not recommend it even if these precautions are observed.

When and where to find it. Commonly on rotting wood of conifers or hardwoods during the summer and fall, but the fruitings likely to attract the mycophagist occur along the sides of country roads, logging roads, or in campgrounds on hard packed soil late in the season. In the Priest Lake district of Idaho, this pattern has been observed repeatedly.

Microscopic characters. **Asci** 200–250 x 12–14 μ. **Spores** 15–18 x 6–8 μ, smooth, narrowly ellipsoid, with 2 oil droplets. **Paraphyses** 200–280 x 4–5 μ(midway) x 7–9 μ(enlarged apex), content dingy yellow brown in water mounts, in KOH becoming cinnabar red in some at least.

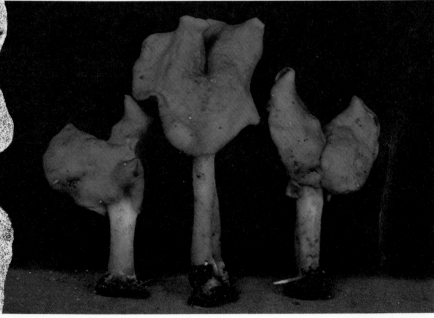

14 *Gyromitra infula*   *About natural size*

Gyromitra esculenta

Field identification marks. (1) Early fruiting period (when the black morels are fruiting); (2) the head at first has a smooth surface but this becomes wrinkled and folded (brain-like in configuration); (3) the stalk (if not compressed) has a single hollow; (4) stalk colored ± like the head.

Observations. This species regularly occurs on humus early in the spring, a feature which aids in distinguishing it from *G. infula.*

Edibility. POISONOUS. The poison, monomethyl-hydrazine, is removed by parboiling (it vaporizes at a lower temperature than water). Parboiling and discarding the water thus makes this an "edible" species. NEVER eat it raw — or any other *Gyromitra* for that matter. Also, organisms have differential tolerances relative to a lethal dose of the poison and the line between no effect at all and a lethal dose can be very thin.

When and where to find it. In conifer forests (and on recent burns?) at the time the black morels fruit — early spring into summer depending on the slope and elevation.

Microscopic characters. **Asci** 8-spored, 280–320 μ long. **Spores** 17–22 x 7–9 μ, smooth. **Paraphyses** 6–8 μ wide near apex, dull cinnabar red when first revived in KOH.

15 *Gyromitra esculenta*   *Slightly less than natural siz*

Hypocreales

Hypomyces

Hypomyces lactifluorum

Field identification marks. It is impossible to identify this mushroom in the field because of its malformed condition.

Observations. The mushroom illustrated has been parasitized. The parasite at the stage shown forms an orange covering over the area where the gills of the mushroom should be. The fruit bodies of the parasite *(Hypomyces)* are small flask shaped structures (perithecia) with a neck the tip of which ends at the surface of the orange layer and is visible with the aid of a hand lens as a slight dot or protrusion technically termed an ostiole. The spores are borne in spore sacs (asci) and when mature are liberated in the perithecium and finally forced out through the neck to the surface and there carried away by air currents.

The aborted mushroom (it never produces any spores of its own) makes up the bulk of this compound fruit body consisting of part an Ascomycete and part a Basidiomycete. There are several common species of *Hypomyces* which attack mushrooms and can be identified in the field by their colors. The one shown is the most common and conspicuous.

Edibility. Edible but not recommended. What if the host mushroom happened to be a poisonous mushroom? Fortunately, apparently, this does not happen very often. The host,

6 *Hypomyces lactifluorum*   *About natural size*

in North America, apparently, is most frequently *Lactarius subvellereus, L. vellereus,* or possibly *L. piperatus* or *Russula brevipes,* but it must be remembered that field identification of the host is practically impossible because of its malformed condition.

When and where to find it. It can be found during the summer and early fall, depending on the season. It is common in hardwood forests east of the Great Plains and common at times in the conifer forests of the Northwest as well as in the hardwood forests of southern Oregon and California.

Microscopic characters. **Asci** cylindric. **Spores** fusiform, apiculate at each end, one-septate, hyaline, roughened when mature, 30–40 x 7–8 μ, white as seen when collected around the ostiole.

Basidiomycetes

Auriculariales

This order features fruit bodies in a relatively unspecialized state of morphological evolution. The distinguishing features are the basidia and the generally gelatinous, cartilaginous, or waxy consistency of the fruit body.

Auricularia

Auricularia auricula   17

Field identification marks. (1) The dark brown color; (2) the rubbery consistency when turgid; (3) the habit of growing on conifer logs especially in the mountains; (4) the earlike to cuplike shape (often depending on the position of the fruit body on the substrate). The critical features are all microscopic.

Observations. The conifer inhabiting strain in the western region may be a distinct species, but regardless of whether it is a "species" or "variety" it serves well to illustrate this odd type of fungus — which in the past was by some mistaken for a true cup-fungus (Ascomycete).

Edibility. In the Orient species of this genus are used for food on a commercial scale and at least one species is imported from the Orient and can be purchased in some American markets. I have no first hand knowledge of the edibility of American variants.

7 *Auricularia auricula*   *About natural size*

When and where to find it. In dense masses or only scattered fruit bodies on rotting fir logs with the bark still on them. This species is common in the mountainous areas of the West after heavy rains in the late summer, fall, or winter.

Microscopic characters. The specimen illustrated was sterile.

Aphyllophorales

This is a very large and complex order composed chiefly of tough to woody species but with a small number popular as edible species. The edible mushrooms are emphasized here.

Key to Families and Odd Genera

1. Spores born on the lining of tubes with their openings (pores) on the underside of the fruit body; fruit body tough to woody or if fleshy the tube layer not ± gelatinous when pressed between the fingers and not separable readily from the context of the cap . . (p. 50) *Polyporaceae*
1. Not as above 2
   2. Spores born on cells covering spines which hang down from the underside of the cap or from a framework of branches . . . . (p. 56) *Hydnaceae*
   2. Spores born on the smooth surface of flattened branches; the fruit body resembling a large bouquet of egg noodles . . . . (p. 62) *Sparassis*

Polyporaceae (Polypores)

The systematics of this large group are still in a state of change with new genera being erected for every group with one or a few fairly distinctive features. For this reason I have used a broad family concept. For the genera, however, I have adapted a more modern point of view for it has been evident for the last 50 years at least that many of the groups of species in the older concepts of the genus *Polyporus* deserved generic recognition. To identify the various genera in this family much microscopic work is required. Only a number of the more distinctive genera are illustrated here.

Key to Genera

1. Fruit body fleshy; stipitate; terrestrial . . . .
   . . . . see (p. 54) *Albatrellus* and (p. 74) *Boletopsis*
1. Fruit body at maturity tough to woody 2

Fomes

Fomes officinalis 18
(Quinine fungus)

Field identification marks. (1) The context is white, cheesy when fresh and brittle when dry; (2) the taste is **very bitter**; (3) the pore surface is white when fresh; (4) the fruit body is hoof-shaped when well developed; (5) it occurs on wood of conifers.

18 *Fomes officinalis*   *About one-third natural size*

Observations. The greenish color is caused by algae grow-ing in the tissues. The fruit bodies most frequently occur on standing tree trunks in which the mycelium is causing a red-dish brown rot. It is quite destructive in old-growth stands in the Rocky Mountains and Pacific Northwest where it attacks various species. It has been harvested for use in making bitters, and as a substitute for quinine.

Edibility. Not edible, see above discussion.

When and where to find it. The fruit bodies are perennial, a new layer of growth being added each year, hence they can be harvested any time during the year. Since they often develop high up on the trunk commercial collectors often used a rifle to dislodge them.

Microscopic characters. **Spores** 4–5 x 3–4 μ, ellipsoid to ovoid, hyaline. **Basidia** 4-spored. **Cystidia** none. **Hyphae of context** long and flexuous, hyaline, unbranched or sparingly branched, lacking cross walls, 2.5–6 μ in diam, KOH on con-text yellowish to pale or dark reddish.

Laetiporus

19 Laetiporus sulphureus var. semialbinus
(Polyporus sulphureus var. semialbinus)

Field identification marks. (1) The white tube layer through-out the development of the fruit body; (2) the lack of yellow tones in the cap; (3) often growing shelving from dead trees or stumps.

Observations. *Polyporus cincinnatus,* with white tubes and a milklike latex in the stalk is a species in its own right not

19 *Laetiporus sulphureus* var. *semialbinus*   *About one-fourth natural size*

to be confused with *L. sulphureus* var. *semialbinus* which lacks a latex. Var. *semialbinus* appears to be a genetically constant variant of wide distribution and one based on the failure of the typical yellow pigment of the type variety to form.

Edibility.   Reported to be as good as var. *sulphureus* — good if collected young enough. Use the soft marginal area of the cap.

When and where to find it.   In the West I have seen it only rarely in southern Oregon in the fall, but voucher specimens from the western area were not obtained.

Microscopic characters.   No description of this variant is included here since the above data are sufficient for an identification and the western material was not saved.

Osteina

Osteina obducta   20

Field identification marks.   (1) The compound fruit body ± resembling that of *Polypilus frondosus;* (2) the pale to dark yellow brown caps; (3) the presence of a gnarled fleshy underground stalklike structure which discolors dull cinnamon when cut; (4) the nondistinctive taste and odor; (5) the white context of the cap which is fleshy-firm when moist but bone-hard when dried; (6) the context staining pale cinnamon with iron salts.

Observations.   The correct name for this species may be *Polyporus zelleri.* The problem needs to be studied.

20 *Osteina obducta*   *About one-half natural size*

Edibility. Not known and hardly worth trying.

When and where to find it. It is usually found in the fall in close proximity to a stump or old conifer tree. Known to date mostly from the Priest Lake district of Idaho.

Microscopic characters. **Spore deposit** white. **Spores** 4.5–6 x 2–2.5 μ, cylindric to oblong, hyaline under the microscope, smooth, inamyloid. **Basidia** 4-spored, 18–26 x 3–4 μ, narrowly clavate. **Pleurocystidia** 5–7 μ wide, absent or very rare and obventricose, otherwise resembling sterile basidia. **Hyphae of the hymenophore** 3–5 μ wide, tubular, hyaline, thin-walled and with looping clamps. **Hyphae of context** very intricately interwoven, hyaline and refractive in KOH, clamped (clamps "normal").

Albatrellus

Hymenophore in the form of tubes, the hymenophoral hyphae parallel to interwoven; fruit body fleshy, growing on the ground in the manner of a bolete; spores smooth, colorless, thin-walled; stalk central to eccentric, more rarely lateral.

Key to Species

1.   Cap white to yellow or in some violaceous gray over the disc (a variant from central Idaho) ......(p. 55) *A. ovinus*
1.   Cap pale tan from the first, often under hemlock ....
.................................(p. 55) *A. confluens*

21 *Albatrellus ovinus*   *About one-half natural size*

Field identification marks.   (1) The white to creamy or yellowish fruit body when it is young; (2) the wavy undulating cap margin; (3) pores 2–4 per mm, white, decurrent on the stalk and often stained yellow; (4) tending to occur in confluent masses.

Observations.   There is need for a critical study of *Albatrellus* in North America. In Idaho, where *A. ovinus* is common, many fruitings are found with caps tinged pinkish gray to pale violaceous brown over almost the entire cap. Pouzar has named this variant *A. avellaneus,* but in the collections I have made, these aberrant fruiting bodies never appeared to be clearly distinguished from the typical variety on other characters. A species with the aspect of *A. ovinus* but with a dull blue cap is *A. flettii,* an endemic species.

Edibility.   Edible, at least so reported, and of possible value as an edible species in the central Idaho area and the Pacific Northwest. In Europe, however, recent reports of intoxication from eating it have been noted.

When and where to find it.   Solitary, gregarious or in confluent masses at times involving twenty-five to fifty fruit bodies. Often abundant after heavy showers in July and August in the Salmon River country of Idaho, but known generally throughout the Northwest.

Microscopic characters.   **Spores** 3.5–4.5 x 2.5–4 μ, ellipsoid to subglobose, smooth, hyaline, inamyloid. **Basidia** 4-spored. **Cystidia** none. **Hyphae of context** thin-walled, the cells inflated, septate but clamps lacking, the hyphae 5–20 μ wide.

Albatrellus confluens   22

Field identification marks.   (1) The tan colored caps when young; (2) the compound fruit bodies (masses of fused caps); (3) the white to yellowish pores; (4) the fleshy consistency of the context.

Observations.   The difference in color between this fungus and *A. ovinus* once seen, is enough to allow the two to be distinguished at sight because the pattern of variability is so different, at least as far as our western *A. ovinus* is concerned.

Edibility.   Not known because it has often been confused with *A. ovinus* which is known to be edible — so the probability is that if it were poisonous we would know by now.

When and where to find it.   Masses of caps occur scattered to gregarious under hemlock. It was seen around Upper Priest Lake in northern Idaho in the fall, but is apparently rare in the West.

Microscopic characters. **Spores** 3–4.5 x 2.5–3 μ, ellipsoid, smooth, hyaline, inamyloid. **Basidia** 4-spored. **Cystidia** none. **Hyphae of context** thin-walled, the cells inflated, 3–9 μ wide, septate but lacking clamps.

22 *Albatrellus confluens*

About one-half natural size

Hydnaceae

The species grouped here, in recent classifications, are divided into two families, the Hericiaceae and Hydnaceae. Fleshy species with amyloid spores and the basidiocarp consisting of a tubercle or a framework of branches are placed in the former.

Key to Genera

1. Fruit body lacking a cap; teeth hanging from a tuber-like mass of tissue or from a system of branches . . . .
   . . . . (p. 58) *Hericium*
1. Fruit body with a distinct cap 2
   2. Fruit body fleshy 3
   2. Fruit body tough; spore deposit white . . . . (p. 57) *Phellodon*
3. Spore deposit white; spores smooth . . . . (p. 61) *Dentinum*
3. Spore deposit dull brown; spores angular, warty (p. 59) *Hydnum*

Phellodon tomentosus   23

Field identification marks. (1) The white spore deposit; (2) the very thin context leathery to semiwoody in texture; (3) the cap surface strongly marked with concentric bands of different color; (4) the margin during the period of rapid growth white, and the remainder dark yellow brown (near "bister" in Ridgway) together are quite definitive.

Observations. The odor though faint is fragrant, and the taste slightly sweetish but leaving a rasping sensation in the throat. These features are not as obvious as those given above but aid in identification. It often occurs in large mats of fused caps almost producing a ceiling over large areas of the moss under conifers.

Edibility. Not edible because of the texture.

When and where to find it. Gregarious to concrescent over areas a foot or more in diameter during good seasons, on moss or needle carpets under conifers. It is common in the forests of western hemlock but not restricted to them. It fruits best during warm wet seasons late in the fall and is, obviously, rather conspicuous.

Microscopic characters. **Spores** white in mass, 3–4 μ diam., globose to subglobose, finely echinulate, apiculate. **Basidia** 4-spored. **Hyphae of pileus** about 4 μ wide, thin-walled, septate; **of stipe** 6.3 μ wide and walls somewhat thickened, rarely branching.

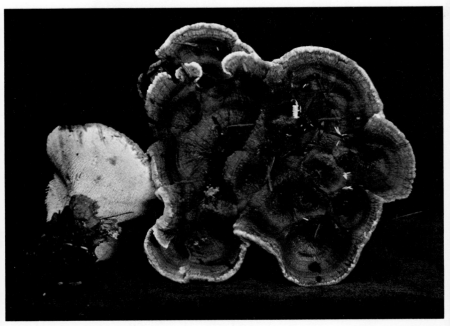

23 *Phellodon tomentosus*   *About natural size*

24   Hericium abietis

Field identification marks.   (1) Fruiting body gigantic (up to 75 cm high); (2) color when young salmon yellow to yellowish orange; (3) the spines long remaining short (about 1 cm long); (4) arising from a fleshy tuber soon divided near the outside into coarse short branches.

Observations.   *Hericium erinaceus* occurs on oak in southern Oregon and in California. It is basically a fleshy unbranched tuber with long spines hanging from it. *H. abietis* occurs on conifer wood and in age is a framework of branches. In *The Mushroom Hunter's Field Guide* it is included under the name *H. weirii.*

Edibility.   Edible and reported as good by a number of collectors. The young fruit bodies when chopped up and cooked are said to make a very acceptable dish. *H. erinaceus* is edible but inclined to be tough and is generally rare in the West.

24 *Hericium abietis*   *Less than one-fourth natural size*

When and where to find it. It occurs on both standing dead trunks and on logs or fallen trees during late summer and fall; not uncommon, especially in the Olympic National Park.

Microscopic characters. Tramal hyphae amyloid, clamped. Spores 4.5–5.5(6.0) x 4.0–4.5(5.0) μ, finely roughened, strongly amyloid.

Hydnum (Hedgehog Mushrooms)

Key to Species

1. Cap, spines, and fleshy context dark violet . . . .
. . . . (p. 59) *H. fuscoindicum*

1. Cap, spines, and context brown; in humid weather exuding a pale yellow juice; stalk fibrous-tough, cap ± fleshy . . . . (p. 60) *H. stereosarcinon*

Hydnum fuscoindicum   **25**

Field identification marks. (1) The dark violet color of the cap and its flesh, the stalk, and the young teeth; (2) the fleshy consistency; (3) the lack of a distinctive odor or taste (or the odor faint and slightly resembling that of cinnamon).

Observations. *Hydnellum regium* may be mistaken for *H. fuscoindicum* but is at once distinguished by the corky brittle pinkish cinnamon stipe and though violaceous tones dominate the cap, they are not the dark indigo of *H. fuscoindicum*.

25 *Hydnum fuscoindicum*   *About natural size*

Edibility. I have few reliable reports on its edibility; apparently it is edible.

When and where to find it. Solitary, gregarious, or in clusters, on duff in conifer woods, especially under pine. It fruits during the late summer and fall and is not uncommon though I have found it more regularly in northern Idaho than anywhere else in the West.

Microscopic characters. **Spores** 5–6.5 x 4.5–5 μ, ellipsoid to subglobose in outline, tuberculate, with 8–10 truncated to rounded warts visible in optical section, appearing weakly amyloid (mount in Melzer's medium). **Basidia** 7–8 μ wide, clavate. **Hymenium** about 35 μ deep. **Context of the spines** of parallel hyphae having thin walls and measuring 5–7 μ wide. **Hyphae of context** of interwoven hyphae having thin walls and with some cells inflated up to 25 μ wide, most hyphae containing granules which in KOH are blue green and darker in Melzer's; oleiferous hyphae present in KOH and these are seen to be filled with a resinous material. **Clamps** absent.

---

26   Hydnum stereosarcinon

Field identification marks. (1) The cap at maturity is a dark rusty brown and finally blackish as the surface covering of the cap begins to collapse from continuous rain; (2) the spines are a rusty cinnamon when mature; (3) the stalk is fibrous-tough; (4) the taste is somewhat farinaceous (like that of fresh meal).

Observations. This species is somewhat intermediate between the genera *Hydnum* and *Hydnellum* since the cap is

26 *Hydnum stereosarcinon*   *About one-half natural size*

truly fleshy at first but the stalk becomes almost woody in age. The disintegration of the skin over the cap in wet weather is interesting as most fungi soften in the interior of the fruit body first. The cluster collected was in a large rosette.

Edibility.   Not edible.

When and where to find it.   Gregarious in rosettes (or clusters) of many caps or in small groups. The collection photographed was made in old-growth hemlock and Sitka spruce late in November. It appears to favor the coastal area.

Microscopic characters.   **Spores** subglobose to oblong, 4.5–5 x 3.5–4.5 μ.

Dentinum

Dentinum repandum   27

Field identification marks.   (1) The spore deposit is white; (2) a cream colored to tan to reddish cinnamon cap; (3) fragile fleshy context; (4) mild taste when fresh; (5) whitish spines on underside of cap.

Observations.   The smooth subglobose spores 6–8 x 5–6 μ are an important microscopic character. In our western states the large nearly white variant is the one most collectors prefer. *D. umbilicatum* also occurs in the region. It

27 *Dentinum repandum*   *About one-half natural size*

features a slender stalk and usually a depression in the center of the cap.

Edibility. Edible and choice.

When and where to find it. Fruiting occurs in the late summer and fall on into early winter in northern California. It is often abundant in the mixed coastal forests but it occurs in the conifer forests of the region generally. In the oak forests the reddish cinnamon variant is more likely to be found.

Microscopic characters. **Spore deposit** white. **Spores** 6–8 x 5–6 μ, subglobose, smooth.

Sparassis

28 Sparassis radicata

Field identification marks. (1) The gigantic clusters give the impression of a large bouquet of leaf lettuce or egg noodles because of the flattened tips of the branches; (2) the pale colors (whitish to yellowish or pale tan); (3) the occurrence in proximity of conifer trees; (4) the consistency is fairly pliant-cartilaginous.

Observations. At the 1972 mushroom show of the Olympia (Washington) Audubon Society, mature clusters were brought in which showed secondary growth from the tips of the branches. Such secondary growth appears to be a rare occurrence in this species.

Edibility. Edible and choice; it is considered a prize by connoisseurs. It is also a safe species for beginners.

When and where to find it. It fruits in the fall in old-growth conifer forests in our western states north into Canada.

Microscopic characters. **Spores** 5–6 x 2.8–3.5 (4) μ, broadly ellipsoid, smooth, inamyloid, hyaline. **Cystidia** none.

28 *Sparassis radicata*

About one-third natural size

Cantharellales

Key to Genera

1. Fruit body ± mushroomlike but gills typically with obtuse edges to veinlike and subdistant to distant .. . .(p. 69) *Cantharellus*
1. Not as above 2
   2. Fruit body much branched and typically large; hymenium smooth; context and/or hymenium giving a grayish blue to olive reaction; the large coral fungi (if branches are flattened see *Sparassis* also) . .(p. 66) *Ramaria*
   2. Fruit body simple or rarely branched once or twice. 3
3. Fruit body an upright club; interior soft and punky; hymenial surface smooth to irregularly wrinkled but not gill-like . . . .(p. 63) *Clavariadelphus*
3. Fruit body tall and slender; soon hollow and moderately pliant; attached to sticks . . . .(p. 65) *Macrotyphula*

Clavariadelphus

Key to Species

1. Fruit body becoming flattened to depressed on apex, 1.5–5 cm thick near apex . . . .(p. 63) *C. borealis*
1. Fruit body usually with a distinct pointed small apical protuberance, 5–15 mm thick near apex (p. 64) *C. mucronatus*

Clavariadelphus borealis   **29**

Field identification marks. This species cannot be distinguished accurately in the field from *Clavariadelphus truncatus*. A spore deposit must be obtained. It is white in *C. borealis* and yellowish in *C. truncatus*.

Edibility. Edible in the sense that it is not known to be poisonous. The disagreeable bitter-sweet taste will not recommend it to most hunters.

When and where to find it. This is the common truncate *Clavariadelphus* of the conifer forests from Oregon to Alaska east to northern Idaho during late summer and fall. It occurs solitary, gregarious, or in small clusters.

Microscopic characters. **Spore deposit** white. **Spores** 9–12 x 4.5–6.5 μ, ellipsoid, smooth, hyaline, with yellowish droplets or granulose content. **Basidia** mostly 4-spored (a few 2-spored). **Cystidia** none. **Apex of club** sterile (scattered basidia only near the margin), in at least some collections covered with a hymeniform layer of clavate to fusoid-ventricose cells. **Clamps** present.

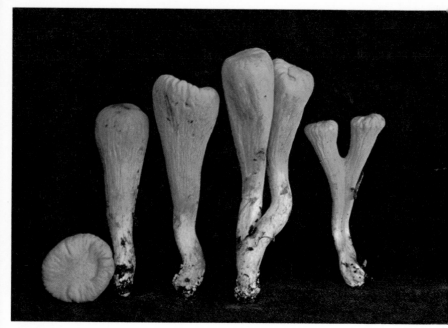

29 *Clavariadelphus borealis*   *Slightly less than natural size*

30   Clavariadelphus mucronatus

Field identification marks.   For this species the identification marks are about as simple as it is possible to make them: Look at the picture and note the sharp point (mucro) on the apex of some of the fruit bodies.

30 *Clavariadelphus mucronatus*   *About natural size*

Observations. As indicated above, the illustration speaks for itself. It has been intimated that the point on the apex is not a constant feature in this genus but is characteristically present to some extent on most species as part of their normal development. From extensive observations on this genus in North America I can state emphatically that this is not true in our populations. In *C. mucronatus,* which, like *C. sachalinensis,* occurs in quantity when it fruits, by far the majority of the fruit bodies show the character.

Edibility. Not tested. The small size of the fruit bodies will discourage most "pot hunters."

When and where to find it. It is known from Alaska to Oregon east to Idaho. Near Upper Priest Lake in Idaho it often fruits in great profusion under *Thuja plicata* (western red cedar) during warm wet fall weather.

Microscopic characters. **Spores** white in deposit, (10.5) 11–15 (16.5) x 3.5–5 μ, in face view narrowly suboblong, in profile slightly inequilateral to "sway-backed," smooth hyaline but with yellow oil drops. **Basidia** 2- and 4-spored. **Cystidia** none. **Clamps** present.

Macrotyphula

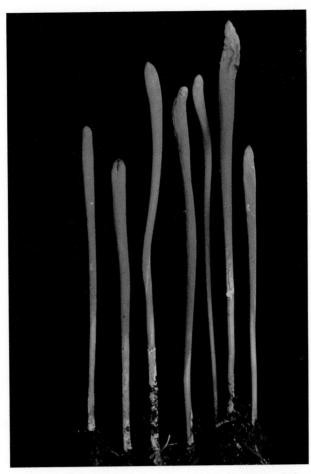

31 *Macrotyphula fistulosa*   *About natural size*

Field identification marks. (1) The exceptionally long slender fruit body; (2) growing on dead sticks and branches, and the base of the stalk having coarse long tawny hairs; (3) consistency relatively cartilaginous; (4) fruit body hollow.

Observations. In this instance I prefer to use the new name rather than the combination with *Clavariadelphus* which Corner (1950) proposed. To me the species is not closely related to the other species of *Clavariadelphus*.

Edibility. Not tested, at least to my knowledge.

When and where to find it. Solitary to scattered on debris (sticks, etc., mainly of alder) in our western area during the fall season after heavy rains. Material for the photograph was found on the southside of Hoodoo Mountain near Priest River, Idaho.

Microscopic characters. **Spores** (12) 14–18 x 7–9 μ, obscurely inequilateral in profile, subelliptic to drop-shaped in face view, smooth, inamyloid. **Basidia** 4-spored. **Cystidia** none. **Clamps** none.

Ramaria

The large conspicuous coral fungi are placed in this genus. In addition to the much-branched feature, the species have dull yellow to buff-colored spores in deposits, and stain green to olive or olive gray on the hymenial surface with iron salts. The hyphae of the fruit bodies are typically thin-walled. The western area is very rich in species of this genus. For a critical account of the genus in western Washington, see Marr and Stuntz, "*Ramaria* of Western Washington," *Bibliotheca Mycologica*. Generally speaking, the large corals are edible, but a few, such as *R. formosa,* cause mild upsets. *R. formosa* has pinkish branches and yellow tips.

Key to Species

1. Fruit body with a white trunk and above it the branches are smoky violet to dark olive brown, violet umber to olive umber ....(p. 67) *R. fennica*
1. Not as above 2
  2. Fruit body amber brown with yellow tips; cut context staining vinaceous brown ....(p. 67) *R. brunnea*
  2. Tips of branches pale yellow; lower branches staining red to vinaceous ....(p. 68) *R. sanguinea*

Field identification marks. (1) The rusty yellow color of the upper branches; (2) the taste is bitter; (3) when cut the exposed surfaces change to vinaceous then to vinaceous brown.

Observations. This is one of Oregon's most beautiful species. As species of *Ramaria* are delimited, I would not consider it close to *R. formosa.*

Edibility. Not recommended because of the bitter taste.

When and where to find it. Solitary to scattered on duff under conifers, especially the true firs *(Abies),* late summer and fall. It occurs throughout the Northwest and as far east as Michigan.

Microscopic characters. **Spores** ochraceous in mass, 10.5–13 x 4.5–5 μ, verrucose, elongate-ellipsoid. **Basidia** 4-spored. **Cystidia** none.

32 *Ramaria brunnea*   *About one-half natural size*

Field identification marks. (1) The dark violet umber to olive umber color over the lower part of the secondary branches; (2) the massive white primary trunk; (3) the tips of the ultimate branches olive yellowish; (4) the slightly bitter taste.

Observations. The above identification is tentative pending a critical restudy of *R. fennica, R. fumigata* and related variants. All in this group are rare during most seasons.

Edibility. Not recommended — it is probably bitter when cooked.

When and where to find it. The specimen illustrated was from the conifer forest along the Upper Priest River in northern Idaho. It fruits during the late summer and early fall.

Microscopic characters. **Spores** 8.5–12 x 3.5–5 μ, ochraceous in deposit, verruculose, ellipsoid, inamyloid. **Basidia** 4-spored. **Cystidia** none. **Clamps** present.

33 *Ramaria fennica*   *About one-half natural siz*

34   Ramaria sanguinea

Field identification marks. (1) The lower branches and base of the mushroom stain vinaceous red where injured or such stains may be evident when the collection is made; (2) the entire branching framework above the primary and secondary branches is yellow with the tips being a brighter yellow; (3) the odor and taste are not distinctive.

Observations. The stains darken from red to dull purplish red (vinaceous). There are many species in the West with more orange or golden yellow colors than this one which is essentially a pale clear yellow.

Edibility. Edible. From all information I have been able to gather it is frequently collected for food.

When and where to find it. This species is abundant under hardwoods in the Great Lakes area, and I have seen it in the fall in southwestern Oregon in mixed forests with oak present.

In contrast to many groups of fleshy fungi, the coral fungi are not as closely restricted to species or genera of seed plants as most gilled fungi, i.e., they may be quite "generalized" in their mycorrhizal-forming ability, or they may be mostly humus-reducing organisms as is the case of such groups of gill-fungi as the *Coprinaceae. R. sanguinea* fruits during the summer and early fall.

Microscopic characters. **Spores** ochraceous in a deposit, 8.5–11 (12.5) x 3.5–4 μ, verrucose-rugulose, oblong to narrowly ellipsoid, inamyloid. **Basidia** 4-spored. **Cystidia** none. **Clamps** absent.

34 *Ramaria sanguinea*   *About natural size*

Cantharellus

Over the years I have given considerable thought to the problem of recognizing *Gomphus* as a genus. In fact I did recognize it at one time — but I have finally concluded that a genus based on a slight pigmentation of the spore wall, especially in the order Cantharellales, is untenable.

Key to Species

---

35   Cantharellus subalbidus

Field identification marks. (1) The very short, stocky sta-ture; (2) the white overall color, but staining rusty yellow to orange and finally orange brown where injured; (3) the very narrow (foldlike) close gills.

Observations. The white spore deposit is the additional character to be checked in order to make the identification final. The fruit body of this species is a model of nature's inefficiency in using raw materials for spore production: The amount of supporting structure relative to the amount of spore producing tissue is the reverse of that for most mush-room fruiting bodies and one reason for regarding this spe-cies (and the genus *Cantharellus* for that matter), as primitive in the scale of evolution.

Edibility. Edible and choice. To the "pot hunter" the fact that the fruit body is so "meaty" is a decided advantage. I regard this as one of the very best of western edible fungi.

35 *Cantharellus subalbidus*   *About one-half natural size*

When and where to find it. Scattered to gregarious during late summer and fall under Douglas fir and under lodgepole pine — as for most chanterelles, it is not highly restricted to one species or genus of trees.

Microscopic characters. **Spore deposit** white. **Spores** 7–9 x 5–5.5 μ, ellipsoid to broadly ellipsoid, smooth, yellow in Melzer's. **Basidia** 62–80 x 8.5–10 μ, 4- to 6-spored. **Cystidia** none. **Pileus** lacking a distinct cuticle, hyphae merely more compactly arranged at the surface. **Clamps** present.

Cantharellus cibarius 36

Field identification marks. (1) The very narrow yellow gills with blunt edges; (2) the wavy-lobed cap margin; (3) the odor of dried peaches (fragrant); (4) the yellowish spore deposit; and (5) the lack of a veil.

Observations. For the purposes of this book a broad species concept is most useful. If one wishes to be highly technical a number of "microspecies" or "subspecies" or "varieties" can be recognized in our western area, but, as far as I am concerned, they seem to intergrade enough to discourage giving names to them. One variant is of particular interest from the standpoint of evolution: At the time of the Stuntz foray at the Cispus Environmental Center near Randle, Washington in 1972, specimens were brought in on which the gills had failed to develop. In one instance two caps, one with gills and the other with a smooth hymenium, came from a common stalk.

Edibility. Edible and choice; also common. It is one of the most popular species in the area.

36 *Cantharellus cibarius*   *Slightly less than natural size*

When and where to find it. In conifer forests generally throughout the area, but often very abundant in pole-size stands of second growth Douglas fir. It fruits during the fall season on into winter.

Microscopic characters. **Spore deposit** pale yellow. **Spores** 7–9 x 4–5 μ, smooth, not amyloid. **Basidia** 50–70 x 6–8 μ, narrowly clavate, mostly 4-spored. **Gill trama** of loosely interwoven hyphae. **Cuticle** of pileus of interwoven hyphae. **Clamps** present.

---

37   Cantharellus kauffmanii

Field identification marks. (1) The lack of orange or red tones in the color of the cap; (2) the dull crust brown straplike scales often covering the "bottom" of the funnel; (3) the merulioid (shallow pored) hymenophore at maturity; (4) the rather fibrous consistency of expanded fruit bodies.

Observations. Old fruit bodies of *C. bonari* may be confused with this species particularly if the color has faded. Some authors, including myself at one time, placed this species in *Gomphus,* but the slight color in the spore wall alone or in combination with other secondary characters hardly is sufficient for the recognition of a genus.

Edibility. The consistency is against this mushroom being a good edible species. In addition, I have no reliable data on tests. In view of the situation with regard to *C. floccosus* it should not be assumed that *C. kauffmanii* is edible.

When and where to find it. Scattered to clustered in old-growth conifer forests in the Pacific Northwest in the fall season. Not common.

37 *Cantharellus kauffmanii*   *About one-half natural size*

Microscopic characters. **Spores** pale ochraceous in thin deposits, 12–15 x 5–7 μ, exospore slightly wrinkled, weakly dextrinoid, subelliptic in face view, in profile with a distinct suprahilar depression. **Basidia** 60–80 x 10–13 μ, 2- and 4-spored. **Cystidia** none. **Clamps** absent.

Cantharellus floccosus 38

Field identification marks. (1) The fruit body hollowed from early stages on to maturity and the hollow, at first, lined with fairly delicate soft scales; (2) configuration of the hymenophore intermediate between gills and shallow pores; (3) the inner surface of the "trumpet" orange red when young, and orange to reddish tints persist.

Observations. The specimens illustrated show the typical form of this species complex. *C. floccosus* is distinct from *C. kauffmanii* in having red to orange pigmentation in the cap and greatly reduced development of the scales. *C. bonari* has very coarse scales, but is ± fire red.

Edibility. This is the variant that has caused the most cases of "mild" mushroom poisoning, though some people can eat and enjoy it with no ill aftereffects.

When and where to find it. In our western region the typical form illustrated here is most likely to be found in conifer forests where hemlock is present. It fruits during the late summer and fall.

Microscopic characters. **Spore deposit** ochraceous. **Spores** 12–15 x 6–7.5 μ, outer wall slightly wrinkled, weakly dextrinoid. **Basidia** 52–60 x 10–12 μ, 4- to 6-spored. **Cystidia** none. **Clamps** absent.

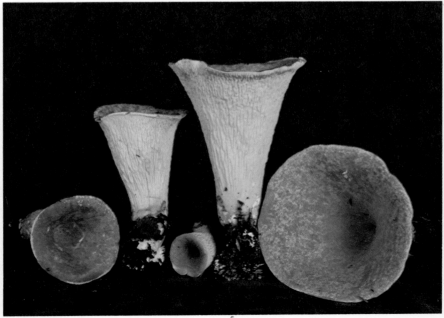

38 *Cantharellus floccosus*   *About one-half natural size*

39   Boletopsis griseus

Field identification marks.   (1) The rather dry-fleshy consistency; (2) the undulating to lobed margin of the cap; (3) the grayish white color at first and the darkening in streaks or patches finally to drab gray; (4) the minute whitish short tubes with fine whitish pores; (5) the short thick stalk.

Observations.   *B. leucomelas* differs by developing a bluish fuscous color over the entire cap. It also occurs in the western area. Both *B. griseus* and *B. leucomelas* have colorless tuberculate spores and tube trama (internal tissues) in which the hyphae do not diverge toward the hymenium. These features distinguish the genus from the *Boletaceae.*

Edibility.   Both species of *Boletopsis* are, apparently, edible but have a bitter taste.

When and where to find it.   Both are frequently encountered in our western states during the fall season. *B. griseus* occurs regularly in the coastal sand dunes under pine, and less regularly throughout the whole western area. *B. leucomelas* is not rare in the Priest Lake area of Idaho, at least during some seasons.

Microscopic characters.   **Spores** 5–6.5 x 3.5–4.5 (5) μ, angular-tuberculate, hyaline in KOH. **Pleurocystidia** none observed. **Clamps** readily demonstrated.

39 *Boletopsis griseus*   *About one-half natural size*

Agaricales

Key to Families

1. Spore producing layer of tubes varying to somewhat gill-like (pore opening extended radially), the layer usually somewhat gelatinous when pressed between the fingers and readily separable from the cap in many species .... (p. 76) *Boletaceae*

1. Spore producing layer in the form of gills 2

   2. Spore deposit gray to smoky brown or olive smoky brown; gills usually decurrent (extending downward on the stalk), thick and often distant; species forming mycorrhiza with conifer trees .... (p. 97) *Gomphidiaceae*

   2. Not as above 3

3. Gills waxy; spore deposit white; basidia usually elongated (45–70 x 7–10 μ) as in *Cantharellus* .... (p. 102) *Hygrophoraceae*

3. Gills not as above; basidia wider in relation to length ($\pm$ 20–40 x 8–14 μ) 4

   4. Spore deposit white to yellowish or pinkish buff; cap and stalk confluent (not cleanly and readily separated from each other) .... (p. 109) *Tricholomataceae*

   4. Not as above 5

5. Spore deposit white, yellow to orange or green 6

5. Spore deposit pink, reddish, reddish cinnamon, rusty brown, purple brown or black 8

   6. Volva absent, outer and inner veils often intergrown; ring typically present on stalk; gills free from stalk .. .. (p. 160) *Lepiotaceae*

   6. Not as above 7

7. Veils lacking and fruit body with a thick short stalk which is rather fragile; if stalk is thin a latex is present (technically this family is defined on the presence of heteromerous fruit-body trama and spores with amyloid ornamentation) .... (p. 235) *Russulaceae*

7. Volva and annulus present, or only a volva present; gills free; or fruit body covered by a slime veil, and the gills free .... (p. 164) *Amanitaceae*

   8. Spores in deposit pink to vinaceous cinnamon 9

   8. Spores in deposit rusty brown, purple brown, or blackish 10

9.   Cap and stalk not cleanly and readily separable; gills attached to the stalk . . . . (p. 171) *Rhodophyllaceae*
9.   Cap and stalk readily separable; gills free . . . . (p. 172) *Volvariaceae*
    10. Spore deposit in crust brown to dark rusty brown series . . . . (p. 173) *Cortinariaceae*
    10. Spore deposit in the chocolate to black series 11
11.   Cap and stalk cleanly separable; gills free; a ring usually present on the stalk . . . . (p. 220) *Agaricaceae*
11.   Not as above 12
    12. Lamellae soon liquefying or if not the caps very rigid — fragile when fresh . . . . (p. 228) *Coprinaceae*
    12. Lamellae never liquefying; caps pliant when fresh, often with a slimy or sticky surface . .(p. 215) *Strophariaceae*

Boletaceae (Fleshy Pore-Fungi)

Only one family is recognized for pore fungi with fleshy readily decaying fruit bodies typically centrally stipitate, and with the hymenophoral context as seen in a longitudinal section having hyphae more or less in bilateral arrangement. The family is placed in the Agaricales, and it should be pointed out that in one genus placed here by some authors, *Phylloporus,* the hymenophore is typically lamellate. I place *Phylloporus* in the Paxillaceae following Watling. An asterisk indicates that no species is included here, but one may expect to find species of these genera in the region. All of these are edible but some are not desirable because of their bitter taste (especially *Strobilomyces* and *Fuscoboletinus).*

Key to Genera

1.   Spore deposit not obtainable — the tubes usually oriented in such a way that the spores cannot fall free from them; spores not discharged from the basidia in the normal manner of species in the Agaricales . . . . *Gastroboletus*
1.   Spore deposit readily obtained from fruit bodies in good condition and near maturity .2
    2. Cap covered with coarse dry gray to blackish scales; tubes white at first, blackening in age; spores spherical to nearly spherical and their surface reticulate to warty; spore deposit blackish brown or nearly so . . .*Strobilomyces*
    2. Cap not as above; spores smooth or if elongated then either smooth or ornamented 3

3.   Spores ornamented by longitudinal wings, folds, or striations . . . . *Boletellus*

3.   Spores smooth or the ornamentation different than in above choice 4

     4.   With any two or more of the following features in combination:

         (a) hymenophore boletinoid at maturity; *(b)* stalk glandular dotted; *(c)* veil leaving a ring on stalk; *(d)* cap slimy to sticky; *(e)* pleurocystidia in bundles and as revived in KOH with incrusting brown pigment around or on the bundle 5

     4.   Not with above combinations of features 6

5.   Spore deposit grayish brown to wood brown, vinaceous brown, purplish brown, chocolate brown to purple drab . . . . *Fuscoboletinus*

5.   Spore deposit dingy yellow to yellow brown, pale cinnamon tan, olive, olive brown or a greenish mustard yellow . . . . (p. 87) *Suillus*

     6.   Spore deposit pale yellow; spores more or less ellipsoid; stalk hollow at maturity . . . . *Gyroporus*

     6.   Not as above 7

7.   Veil dry and flocculent to almost powdery (in appearance) sulphur yellow and typically leaving a zone or ring on the stalk when it breaks . . . . *Pulveroboletus*

7.   Veil lacking or not as above if one is present 8

     8.   Spore deposit gray brown, red brown, vinaceous, vinaceous brown or purple brown (if stalk bears dark colored points or squamules [small scales] see *Leccinum)* . . . . (p. 94) *Tylopilus*

     8.   Spore deposit yellow brown, rusty, yellow, olive, olive brown, dark cinnamon brown to pale cinnamon or amber brown 9

9.   Stalk scabrous-roughened with the ornamentation dark colored from the first or darkening as the fruit bodies age (often blackish finally) . . . . *Leccinum*

9.   Stalk not ornamented as above . . . . (p. 78) *Boletus*

*For accounts and illustrations of *Gastroboletus, Strobilomyces, Boletellus, Fuscoboletinus, Gyroporus, Pulveroboletus,* and *Leccinum* see Alexander H. Smith and Harry D. Thiers, *The Boletes of Michigan* (Ann Arbor: The University of Michigan Press).

Boletus (Boletes)

Key to Species

---

40   Boletus satanus

Field identification marks.   (1) The prominent bulb of button stages — which may be 6 to 12 cm broad, and livid pink to rose color on its margin or upper surface; (2) the pallid dry to subviscid cap when young but becoming pinkish along the margin later; (3) rose pink small pores; (4) the cut surface of the context yellow at first but staining blue near the tube-line, and olive yellow near the cuticle (surface layer).

Observations.   Assuming that *B. satanoides* is not identical with *B. eastwoodiae*, there are three closely related species in the stirps *Satanus; B. satanus*, *B. satanoides*, and *B. eastwoodiae*. Two occur in California. In *B. eastwoodiae* the bulb is much less prominent (its stalk may be cylindric) than in *B. satanus*, the cap is much darker in color, and the red colors are darker and duller — more maroon in tone.

Edibility. Poisonous. Refrain from eating any somewhat similar species — but I have received reports that *B. satanus* is eaten in the San Francisco area.

When and where to find it. Solitary to scattered, mostly under oak in California during the winter (or late fall) season, not common.

Microscopic characters. **Spores** 12–15 x 4–6 μ (17.6 x 8 μ), subelliptic to subcylindric in face view, in profile obscurely inequilateral, ochraceous in Melzer's, smooth. **Basidia** 4-spored. **Cystidia** 28–42 x 7–12 μ, deeply embedded, ochraceous in age, clavate, on sides and edges of tubes. **Clamps** none.

40 *Boletus satanus*   *About two-thirds natural size*

Boletus aereus   41

Field identification marks. (1) The cap is blackish to dark date brown at first, slowly fading unevenly or in local areas to paler brown; (2) the cap is uneven and with a hoary bloom at first; (3) the netting of the stalk becomes brownish in age; (4) the general aspect is that of *B. edulis;* (5) the tissue of the tube layer gives a "fleeting amyloid" reaction when fresh.

Observations. At long last the presence of this species in North America has been established on a reasonably convincing basis (Thiers, in press). It is not, as Smith and Thiers (1971) pointed out, the same as *B. variipes* or *B. atkinsonii.* The best character for a quick distinction is the blackish brown cap.

Edibility. Edible and rated by many as the best edible bolete in California — where the competition is with the best of the wild mushrooms. My advice to collectors is to look in the oak areas of the state in the fall soon after the rains begin.

When and where to find it.   It grows scattered to gregarious, frequently under oak in the fall.

Microscopic characters.   **Spores** 13–16 x 4–5.5 μ, somewhat inequilateral in profile, subfusiform in face view, smooth. **Basidia** 4-spored. **Pleurocystidia** 30–44 x 8–14 μ, fusoid-ventricose. **Cuticle** of pileus of hyphae 5–8 μ wide, in a collapsing turf. **Clamps** apparently absent.

41 *Boletus aereus*   *About two-thirds natural size*

42 *Boletus calopus*   *About one-half natural size*

Field identification marks.   (1) The netted stalk (the netting very fine); (2) red tints somewhere over the base of the stalk or higher; (3) the bitter taste of the raw flesh; (4) bluish transverse septa in the subhymenial hyphae; (5) the elements of the cap's epicutis incrusted and some of the incrustation amyloid. (Numbers 4 and 5 are not field characters but are essential to a correct identification.)

Observations.   The degree to which the red coloration shows on the stalk varies between collections from very little to very pronounced. The caps become very cracked, especially in relatively dry seasons. *B. coniferarum* is a second species showing the amyloid septa in the subhymenial hyphae but it lacks red tints in the stalk and has a darker olive smoky brown cap as a rule.

Edibility.   Not edible: very bitter.

When and where to find it.   Scattered to gregarious in the old-growth forests of the Pacific Northwest in particular, but likely to be found in the northern conifer belt across the continent during the summer and fall; not uncommon at times.

Microscopic characters. Spores olive brown in deposit; 13–19 x 5–6 μ, smooth. **Pleurocystidia** 40–60 x 9–13 μ, fusoid-ventricose, thin-walled, content yellowish in KOH, the same in Melzer's. **Cheilocystidia** clavate to fusoid and strongly ochraceous in KOH or Melzer's. **Tube trama** of more or less parallel hyphae diverging to the subhymenium, hyaline to yellow in KOH or Melzer's, subhymenium indistinct but the septa of the hyphae mostly blue in Melzer's. **Trichoderm** of the pileus of hyphae with dull yellow brown incrustations as seen in KOH mounts, the end cells tubular to cystidioid and also with incrustations, in Melzer's the hyphae of the trichoderm incrusted and the hyphal walls mostly amyloid (the layer blackish in sections mounted in Melzer's). **Hyphae of the subcutis and context** with pale to dark orange brown content in Melzer's and the septa showing as amyloid rings.

Boletus subglabripes   43

Field identification marks.   (1) A moist but not viscid cap which is dull yellow to reddish orange or at times cinnamon to reddish cinnamon, yellow tubes, and a yellow scurfy stalk; (2) no significant change to blue; (3) more frequently associated with birch than any other tree in the Pacific Northwest.

Observations.   *B. subglabripes* was placed in *Leccinum* by Singer but if that genus is limited to species in which the scurf of the stalk darkens in color, then, as Smith and Thiers pointed out, it is logical to retain the species in *Boletus* where it is found to have close relatives.

Edibility.   Reported as edible.

When and where to find it.   To date I have observed it rarely in the northwest and only in northern Idaho (Bonner County) where it is in the alder-birch stands along with such species as *Cortinarius armillatus* and *C. pholideus* during late summer or early fall. During some favorable seasons I expect that it will be found in great quantity, as has now happened with *Amanita phalloides* in California.

Microscopic characters. **Spore deposit** olive to olive brown. **Spores** 11–14 x 3–5 μ, smooth, narrowly fusoid in face view, in profile somewhat inequilateral, pale greenish yellow in KOH, pale yellow in Melzer's, lacking an apical pore. **Pleurocystidia** 32–54 x 8–15 μ, rare to scattered, fusoid-ventricose, thin-walled, smooth, apex acute, content (revived in KOH) hyaline to yellowish. **Cheilocystidia** 20–32 x 8–12 μ, fusoid to fusoid-ventricose, hyaline to yellowish in KOH. **Caulocystidia** clavate to subglobose, 35–60 x 10–30 μ, thin-walled, smooth, soon becoming hyaline in KOH. **Pileus cuticle** a compact trichoderm with the distal 2–5 cells inflated and compacted to produce a cellular layer, the cells 10–25 μ wide and more or less globose, walls yellow revived in KOH. **Clamps** absent from all hyphae as far as observed to date.

43 *Boletus subglabripes*   About natural siz

44   Boletus flaviporus

Field identification marks.   (1) The slimy-sticky cap; (2) its reddish brown to cinnamon chestnut color; (3) the yellowish stalk; (4) the persistently bright yellow tubes; (5) the short "root" (pseudorhiza).

Observations.   Because of the slimy cap in wet weather one's first thought on seeing the species is that it is a *Suillus*, but this is not substantiated by the more fundamental characters of cystidia and spores. Singer described the stalk as concolorous with the cap, and illustrated very robust fruit bodies. He may have been dealing with a different species.

Edibility.   Apparently edible, but I have no data on it.

When and where to find it.   Gregarious under oak and manzanita in California and Oregon, usually soon after the fall rains start, abundant in some localities.

Microscopic characters.   **Spores** 14–17 x 5–6 μ, smooth, somewhat inequilateral in profile, subfusoid in face view, golden yellow in KOH, in Melzer's duller (possibly weakly amyloid) apical pore not distinct. **Basidia** 4-spored. **Pleurocystidia** 38–57 x 10–20 μ, broadly fusoid-ventricose with short necks and obtuse to subacute apex, thin-walled, content hyaline in KOH, smooth. **Cheilocystidia** smooth, 32–47 x 9–17 μ, pedicellate-ovate to clavate-submucronate, on a slender long pedicel, hyaline revived in KOH and thin-walled, smooth. **Pileus cuticle** a gelatinous trichoderm of hyphae 3–5 μ wide, hyaline in KOH, end cells nearly tubular to slightly cystidioid, some cells with granular content. **Clamps** absent.

Boletus flaviporus   *About natural size*

Boletus porosporus var. americanus   45

Field identification marks.   (1) Dry velvety cap when young and becoming areolate-rimose in age; (2) exposed context in the cracks pallid to yellow; (3) color of cap olive to olive fuscous young, olive gray in age; (4) context and pores turn blue readily when injured; (5) stalk not entirely red (or rarely so) and not netted.

Observations.   The spores are truncate — which immediately distinguishes it from *B. chrysenteron,* and little or no red tones develop on the cap. It is readily distinguished from *B. zelleri* by the truncate spores, rimose cap and pores which stain blue readily. *B. truncatus* develops strong red tones along the edge of the cap and in the cracks. Actually this western variant is intermediate between *B. porosporus* of Europe and *B. truncatus.*

Edibility.   Edible and collected by many mushroom hunters.

When and where to find it.   It occurs throughout the western United States but is especially common early in the fall season in the rain forests of the coast. I have found it abundantly during rather dry years.

Microscopic characters.   **Spore deposit** dark olive. **Spores** 11–14 (16) x 4–5.5 μ, smooth, amyloid at first but quickly changing to bister, boletoid, truncate at the apex. **Basidia** 4-spored. **Pleurocystidia** scattered, fusoid-ventricose, 36–58 x 9–14 μ, thin-walled, apex subacute, content hyaline in KOH. **Caulocystidia** with yellow content in KOH when fresh, fusoid, up to 60 μ long and 13 μ wide, smooth, wall up to 0.5 μ thick in some (revived in KOH). **Cuticle** of pileus a well-formed trichoderm mostly of relatively uninflated cells and apical cells rounded or tapered to an obtuse point, wall heavily incrusted with plates of ochraceous material (fresh material mounted in KOH). **Clamps** none observed.

46   Boletus rubripes

Field identification marks.   (1) Taste bitter; (2) stalk not netted; (3) stalk with red areas variously distributed; (4) cap velvety and olive buff but becoming areolate (cracked) in age with the areolae gray brown to dingy yellow brown; (5) the rapid change to deep blue when the context is cut; (6) the robust stature.

Observations.   This species is easily mistaken for *B. calopus* but lacks netting on the stalk. *B. coniferarum* lacks red tints on the stalk, but has a fine netting (use a hand lens to see it). Microscopically *B. rubripes* is readily distinguished because of the lack of amyloid walls in the hyphae of the tube and cap tissues.

Edibility.   Not edible because of the bitter taste. I have no accurate data as to whether or not it is poisonous.

When and where to find it.   It usually occurs solitary on road banks or in wet conifer woods in the Pacific Northwest including Idaho. It appears to be a late summer and fall species often found in company with *B. coniferarum* and *B. calopus.*

Microscopic characters.   **Spores** 12.5–17.6 x 4–5 μ, ochraceous in KOH, dark ochraceous in Melzer's, in face view narrowly elliptic, in profile subfusoid, smooth. **Basidia** 4-spored. **Pileus trama** interwoven; cuticle differentiated as a layer of interwoven hyphae brown to pale ochraceous in KOH, some of them incrusted. **Caulocystidia** in the form of hyphal tips, septate and with relatively thick walls. **Clamps** not present.

5 *Boletus porosporus* var. *americanus*　　　　　　　　　　About two-thirds natural size

6 *Boletus rubripes*　　　　　　　　　　About one-half natural size

Field identification marks. (1) Pores yellow and staining blue; (2) cap at first more or less coated with yellowish olive to olive buff fibrils but developing a strong rose red ground color and in age or when wet often dark rose red; (3) splashes or zones of purple red to rose red on the stalk; (4) typically the base of the stalk narrowed to a blunt point.

Observations. This species is the chameleon of the western boletes because one seldom finds two fruit bodies with exactly the same proportions of the colors emphasized above. However, it is one of the easiest boletes in the west to identify at sight.

Edibility. I have no accurate data on its edibility, and do not recommend amateur collectors experimenting with it. There are just enough poisonous species in *Boletus* which stain blue to indicate caution.

When and where to find it. In Idaho it fruits in late summer and west of the Cascades mostly during the fall–winter season. It is usually found in the spruce-fir zone of Idaho and the spruce-hemlock zone in the coastal area. It is frequently found on the banks of road cuts through such areas and frequently accompanies *B. calopus, B. coniferarum* and *B. subtomentosus.*

Microscopic characters. **Spores** 14–19 x 4–6 μ, smooth, inamyloid, pale ochraceous in KOH, subfusoid in profile view. **Basidia** 4-spored. **Pleurocystidia** 30–60 x 7–12 μ, deeply embedded, hyaline, clavate to fusoid-ventricose. **Pileus trama** interwoven, homogeneous; cuticle differentiated as a layer of loosely interwoven hyphae, when young the hyphal tips often appearing cystidioid and fasciculate, collapsing in age, pale ochraceous in KOH. **Clamps** absent.

47 *Boletus smithii*

About two-thirds natural size

Suillus

For generic characters, see key to genera. This is the most abundant genus of the family in our western conifer forests.

Key to Species

1. Cap typically dry to the touch and with fibrillose scales; growing under Douglas fir . . . . (p. 87) *S. lakei*
1. Not as above 2
    2. Stalk distinctly glandular dotted 3
    2. Stalk not distinctly glandular dotted 5
3. Cap glabrous; ring on stalk a broad band at the time the veil breaks from the cap margin and having a slimy outer layer . . . . (p. 88) *S. subolivaceus*
3. Not as above 4
    4. Cap with brown spotlike scales near margin; ring rarely present . . . . (p. 89) *S. sibiricus*
    4. Cap glabrous (without scales) and olive-colored as it reaches maturity; stalk never with a ring . . . (p. 91) *S. pungens*
5. Stalk with an evanescent ring slimy on its outer side; stalk base staining green when cut and at times green stains developing on the cap surface . . . .
. . . . (p. 91) *S. imitatus* var. *viridescens*
5. False veil forming a marginal cottony zone on cap margin, the underside of the veil with a lilac brown to chocolate-colored zone . . . . (p. 93) *S. borealis*

Suillus lakei   48

Field identification marks. (1) The distinct small fibrillose scales on the cap; (2) the dry ring or some remains of one near the apex of the stalk; (3) when sectioned, the base of the stalk typically stains pea green.

Observations. The name *Suillus amabilis* should not be used in place of *S. lakei* for the species illustrated here. The situation relative to the use of these names and the circumscription of the species to which they apply has been so much a matter of controversy that it cannot be dealt with here. *S. amabilis* is not treated in this work simply because it is still known only from the part of a Bartholomew collection preserved at Harvard. *S. cavipes* resembles *S. lakei* somewhat but grows under larch and has a hollow stalk. It is edible also.

Edibility. Edible. It is used by many collectors, but I have no data on its quality.

48 *Suillus lakei*

When and where to find it. Solitary to gregarious under Douglas fir, apparently throughout its range. It is one of the first species to fruit after the fall rains begin, and in some years can still be found after nearly all other mushrooms have disappeared.

Microscopic characters. Spore deposit dingy cinnamon. **Spores** (7) 8–10 (11) x 3–3.7 (4) μ, in face view subelliptic, in profile somewhat inequilateral, smooth. **Basidia** 4-spored. **Pleurocystidia** and cheilocystidia abundant, as revived in KOH with brown incrusting pigment and/or this also surrounding the bundle when the cystidia are clustered. **Pileus cuticle** a layer of gelatinous interwoven hyphae with an epicutis of incrusted to smooth hyphae brownish ochraceous in KOH, the scales of the pileus formed by erect groups of hyphal tips. **Clamps** present but rare.

49   Suillus subolivaceus

Field identification marks. (1) A conspicuously glandular dotted stalk above and below the ring; (2) a broad ring (at the time the veil breaks) and with the lower edge flanged outward slightly; (3) outer layer of ring slimy; (4) the stalk 10 to 20 mm thick; (5) the cap brownish olivaceous to dingy yellow brown.

Observations. *S. luteus* has a more vinaceous brown cap and a differently colored outer layer on its ring. *S. subluteus* is a smaller bolete with an even more baggy veil before the ring is formed. Actually it is also distinct in the color of the

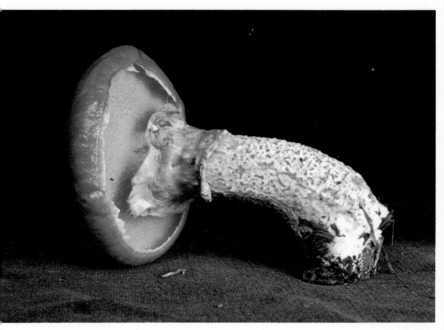

49 *Suillus subolivaceus*   About natural size

interior of the stalk. *S. subolivaceus* is the largest species of the *S. acidus* group.

Edibility.   Not tested and not recommended.

When and where to find it.   Solitary to gregarious under white pine, northern Idaho, late summer and fall during rainy seasons.

Microscopic characters.   **Spore deposit** dingy cinnamon. **Spores** (8) 9–11 x 3–4 (4.4) μ, in face view typically subfusoid, smooth, pale ochraceous to greenish hyaline in KOH. **Basidia** 4-spored. **Pleurocystidia** in fascicles, with rusty brown pigment as revived in KOH; the cells 40–55 x 6–11 μ, subcylindric to narrowly clavate or subfusoid. **Cheilocystidia** as for pleurocystidia. **Epicutis of pileus** a thick layer of gelatinous hyphae 3–7 μ wide and with some bister incrusting pigment (as revived in KOH), the hyphae crooked and appressed. **Clamps** none.

Suillus sibiricus   50

Field identification marks.   (1) The margin of the cap is at first hung with coarse cottony patches of yellowish veil material — especially in those fruit bodies on which a ring does not form; (2) the stalk is typically 1 cm or more thick at the apex; (3) it usually stains dingy vinaceous around the base after handling and tends to become hollow from the collapsed pith in age; (4) a ring is frequently present on robust specimens.

Observations.   In our western area where 5-needle (white) pines occur, this species replaces *S. americanus* of the central and eastern states. The two grade into each other as far as field characters go, but I have not seen a typical fruiting of *S. americanus* in the West. The mottled pattern of veil material over the cap varies with the weather — the veil remnants become washed off as the result of heavy rain. The yellow pigment in the cap is at least somewhat soluble in water causing the color of the cap to vary with the amount of rain that has fallen on it.

50 *Suillus sibiricus*   About two-thirds natural si,

51 *Suillus pungens*   About two-thirds natural s⯑

Edibility. Not recommended because of its thin, usually watery flesh. I have never tried it for flavor. In the button stages, after the gelatinous skin over the cap is removed, there is very little left to eat.

When and where to find it. Scattered to gregarious or the clusters gregarious, appearing in the summer and fall in the northern part of the western states, particularly northern Idaho where white pine is abundant. Once started it may fruit for six weeks in one location.

Microscopic characters. **Spore deposit** dull cinnamon after moisture escapes. **Spores** 8–11 x 3.8–4 μ, narrowly elliptic in face view, narrowly inequilateral in profile, smooth. **Basidia** 4-spored. **Pleurocystidia** 40–70 x 6–9 μ, in clusters, as revived in KOH the clusters surrounded by brown pigment, the cells cylindric to narrowly clavate and often crooked. **Cheilo-cystidia** more or less similar to the pleurocystidia. **Pileus** with a pellicle of appressed gelatinous hyphae 3–6 μ wide and ochraceous in KOH. **Clamps** absent.

Suillus pungens   51

Field identification marks. (1) The color pattern of the cap: white to olive to olive brown to olive yellow; (2) the thick slime layer over the cap (much as in *S. brevipes*); (3) glandular dotted stalk; (4) very weak development of the false veil.

Observations. In the original description not enough emphasis was placed on the olive green phase of the color pattern of the cap — a striking feature. Too much evidence was placed on the development of the false veil — which is typically weak and does not become membranous or leave membranous patches on the cap margin as in *S. albidipes*.

Edibility. Not recommended because of its harsh taste.

When and where to find it. Common under Monterey pine *(Pinus radiata)* in the San Francisco Bay region at the beginning of the mushroom season usually in December.

Microscopic characters. **Spores** 9.5–10 x 2.8–3.5 μ, nearly hyaline in KOH, ellipsoid to subcylindric, smooth, thin-walled. **Basidia** 4-spored. **Pleurocystidia** rare to scattered, usually in large clusters near the pores, the cells 43–79 x 7–10 μ, cylindric to subclavate, as revived in KOH dark brown and with incrusting pigment around the bundle. **Cheilocystidia** similar to pleurocystidia. **Pileus cuticle** a layer of interwoven gelatinized hypae 4–5 μ wide, staining brown in KOH. **Clamps** present.

Suillus imitatus var. viridescens   52

Field identification marks. Buttons can be readily identified by: (1) The slimy glabrous reddish, orange reddish, rusty orange or orange ochraceous caps, the colors becoming paler in the sequence given. Old caps are often merely yellowish to grayish ochraceous; (2) the thick stalk usually pointed at the base; (3) the orange to orange buff band of slime as an outer layer on the ring; and (4) the large pores decurrent on the stalk as a netting.

Observations. In age, many caps will show olive to green areas or almost the entire cap may be so colored or even dark bluish green. This all takes place in perfectly healthy fruit bodies. Also, in warm weather if the base of the stalk is cut, it stains greenish to bluish green in places. If the cut surface of the stalk in the base stains a distinct blue, and the ring does not have an outer gelatinous layer, one most likely has *S. caerulescens.*

Edibility. Edible but not desirable because of the slime. Reports on it indicate inferior to mediocre flavor. There is some danger in accepting reports on this species because it is so easily confused with *S. caerulescens.*

When and where to find it. Cespitose-gregarious to scattered under Douglas fir, hemlock, and lodgepole pine. On the basis of present information it appears to be relatively unrestricted as to tree-associate. It fruits in large quantities along the coast during late summer and fall and is abundant during seasons when mushrooms are generally scarce.

Microscopic characters. **Spore deposit** dingy yellowish brown (but pores stain white paper vinaceous brown). **Spores** 8–10 (12) x 3.8–5 μ, in face view elliptic to subfusoid, in profile somewhat inequilateral, nonamyloid. **Basidia** 4-spored. **Pleurocystidia** solitary as well as in clusters, with dark yellow brown pigment in and around the base of the cluster as revived in KOH; individual cystidia 29–70 x 5–9 μ, cylindric, clavate, or subfusoid. **Cheilocystidia** similar to pleurocystidia. **Caulocystidia** in clusters, the cells 50–70 x 6–12 μ. **Tube trama** of gelatinous somewhat divergent hyphae 6–8 μ wide. **Epicutis of pileus** a gelatinous layer of hyphae 2–4 μ wide, the branching not infrequent.

52 *Suillus imitatus* var. *viridescens*   *About two-thirds natural s.*

Field identification marks. (1) The lilac brown to chocolate colored patches or a zone on the underside of the veil; (2) the fact that the latter separates cleanly from the stalk thus not leaving a ring; (3) that it is a late-season species; (4) that it has the tendency to develop reddish vinaceous colors in the lower portion of the stalk where the latter has been damaged are, taken together, diagnostic.

Observations. *S. luteus* (Slippery Jack) is close to *S. borealis* but regularly has a ring on the stalk and the latter is conspicuously glandular dotted. *S. brunnescens* has a cap which remains white up to early maturity and then becomes chocolate brown, also, it is associated with sugar pine, and the outer layer of the veil does not gelatinize.

Edibility. Choice. We have tried this species on a number of occasions and, when it is collected in the young state, regard it as the equal of any of the highly rated species. The slime of the cap does not wipe off readily so it is best to peel the outer layer at the time of collection and put the clean context which remains in a clean container. The stalk cooks up well in young specimens but in the base one often finds insect larvae.

When and where to find it. It is characteristic to find this species fruiting in quantity late in the fall along roads which pass close to woodlots containing *Pinus monticola* (Western White Pine). It is often on hard packed soil, the stalk being so short the cap appears to be resting on the ground. In northern Idaho, where it is most abundant, it is characteristically a late species, often continuing to fruit after the first series of hard frosts. On a cold wet day it is a rather

3 *Suillus borealis*   *About natural size*

disagreeable fungus to handle, but do not avoid it on this account.

Microscopic characters. **Spores** 7–8 x 2.58–3 μ, narrowly oblong in face view, slightly inequilateral in profile, smooth, yellowish in KOH. **Basidia** 4-spored. **Pleurocystidia** clustered, 38–40 (50) x 8–12 μ, subfusoid, clavate or cylindric, the content and the area surrounding the cystidia vinaceous red in KOH on fresh specimens. **Pileus** with an epicutis of hyphae 4–9 μ wide, gelatinous and appressed to the surface. **Clamps** none.

Tylopilus

Key to Species

54   Tylopilus amylosporus

54 *Tylopilus amylosporus*   *About two-thirds natural siz*

Field identification marks. (1) The cap is olive fuscous to olive brown becoming dingy cinnamon and areolate on aging; (2) the context is reddish under the cuticle, yellow elsewhere, and staining blue when cut; (3) tubes yellowish and the pores stain greenish if bruised; (4) stalk is not netted.

Observations. This species is close to *Boletus porosporus* but is not as olive, the tubes are a duller yellow (to brownish yellow), there is not as much red in the stalk, and the spores are amyloid.

Edibility. Not tested.

When and where to find it. Gregarious under alder, northern Idaho, "rare," in September. It very likely has been overlooked previously.

Microscopic characters. Spore deposit dark wood brown (grayish chocolate color). **Spores** 12–17 x 4.4–6 μ, with a "truncate" apex, in face view subfusoid, in profile somewhat inequilateral, thick-walled, amyloid in crushed mounts of hymenophore. **Basidia** 4-spored. **Pleurocystidia** 40–60 x 8–12 μ, fusoid-ventricose, apex subacute, hyaline in KOH. **Cuticle** of pileus a trichoderm of hyphae 8–15 μ wide, with plate-like incrustations over the surface that are pale bister in KOH; end cells somewhat cystidioid. **Clamps** rare.

Tylopilus olivaceobrunneus   55

Field identification marks. (1) The dark blackish brown to olive brown dry cap; (2) the strongly netted stalk; (3) the lack of blue stains when injured; (4) the dark coffee bean brown pores in youth as well as at maturity are diagnostic.

Tylopilus olivaceobrunneus   *About natural size*

Observations. Previously I had referred to a member of the *Boletus edulis* group under this name. For a comparison with *T. pseudoscaber* see *B. edulis.*

Edibility. Probably edible. I have no data on it in the sense of the original concept.

When and where to find it. Solitary to gregarious under Sitka spruce, during the fall season in Oregon. Apparently it is rare, but this needs verification — it is certainly rare in the sense that it was rarely recognized.

Microscopic characters. **Spores** 13–16 x 5–6 μ, brownish in KOH, smooth, somewhat inequilateral in profile, apex not truncate. **Basidia** 4-spored. **Pleurocystidia** scattered as fusoid-ventricose brown cells (in KOH). **Pileus cuticle** a trichoderm, the end cells cystidioid, the walls smooth or nearly so, cell content dull brown in KOH. **Clamps** none.

56   Tylopilus pseudoscaber

Field identification marks. (1) The dark coffee brown pores which are the outstanding feature; (2) the dark brown dry cap; (3) the dull brown to coffee brown pruinose stalk lacking any netting; (4) the change to blue and then slowly to dark brown on the injured context; (5) the staining of waxed paper blue when this is used as a wrap for fresh material; (6) the more or less chestnut brown spore deposit.

Observations. *T. porphyrosporus* and *T. amylosporus* are both closely related to *T. pseudoscaber* but readily distinguished; the former by having gray pores when fresh, and the latter by its yellowish flesh in the cap. A third species and one having coffee brown pores like *T. pseudoscaber,* is *T. olivaceobrunneus,* but it is readily distinguished by the distinctly netted stalk.

Edibility. Because of previous confusion in the literature in regard to the identity of the above mentioned species, we have no accurate information as to their edibility, though they are probably edible.

When and where to find it. *T. pseudoscaber,* at least during some seasons, is a common fungus along the Pacific Coast, but it also occurs in northern Idaho. It is solitary to gregarious, and does not occur in quantity as do many species of *Suillus,* for instance. I have found it through late summer on into late fall.

Microscopic characters. **Spore deposit** distinctly reddish brown ("chestnut"). **Spores** (12) 14–18 x (5) 6–7.5 μ, occasional giant spores up to 27 μ long present; in face view narrowly ellipsoid, in profile obscurely inequilateral with a shallow suprahilar depression, pale earth brown in KOH, smooth, near "tawny" or more reddish in Melzer's. **Basidia** 4-spored. **Pleurocystidia** scattered, 50–63 x 10–16 μ, clavate-mucronate to fusoid-ventricose, usually with a dark brown content when fresh. **Cheilocystidia** similar to pleurocystidia or up to 30 μ broad, brown when fresh. **Hyphae** of the pileus cuticle 7–9 μ broad, with elliptic to subclavate end cells having the tips tapered to an acute to subacute apex. **Clamps** absent.

About natural size

Gomphidiaceae

This family features the hymenophore of thick usually distinct gills, the olive gray to fuscous spore deposit, the formation of mycorrhizae with conifers, and large "boletoid" spores. Two genera are common in the West, *Gomphidius* and *Chroogomphus*.

Key to Species

1. Flesh white or whitish (except for lower part of stalk where it is lemon yellow) 2
1. Flesh ochraceous to ochraceous buff throughout 4
    2. Veil absent; spores 17—23 μ long; stalk readily staining black . . . . (p. 98) *G. maculatus*
    2. Veil present on young fruit bodies .3
3. Gills close; growing in clusters; spores 10–13 μ long . . . . (p. 98) *G. oregonensis*
3. Gills ± distant; typically gregarious; spores 15–21 x 4–6 μ . . . . (p. 100) *G. glutinosus*
    4. Cap dry, fibrillose, ± ochraceous . . . . (p. 101) *Chroogomphus tomentosus*
    4. Cap viscid at first, soon becoming olive brown to gray brown . . . . (p. 101) *Chroogomphus rutilus*

57 Gomphidius maculatus

Field identification marks. (1) Lack of both a slime veil and a dry fibrillose veil; (2) the tendency of the fruit body to blacken readily; (3) glandular spots over the apical region of the stalk (use a hand lens); (4) the occurrence of the fruit bodies near larch trees.

Observations. For an accurate identification, the spore size should also be taken into account. In a number of species the slime veil tends to become obliterated — so young fruit bodies should be examined.

Edibility. Edible. In northern Idaho it can be collected in sufficient quantity to be tempting as an edible species.

When and where to find it. Solitary or gregarious in the fall in areas where larch is growing. According to my experience it is more abundant around Priest Lake in northern Idaho than anywhere else in North America.

Microscopic characters. **Spores** 17–23 (25) x 6–8 (9) μ, smooth, subfusoid, dark drab in a deposit. **Basidia** 4-spored. **Pleurocystidia** 100–130 (216) x 12–20 (26) μ, ventricose near base, apex obtuse, some with yellow brown incrustations, thin-walled. **Caulocystidia** in patches, similar to pleurocystidia or some thick-walled. **Cuticle** of pileus a gelatinous pellicle of hyphae 2.5–4 μ wide, in crushed mounts in KOH colored deep plumbeous. **Clamps** rare.

58 Gomphidius oregonensis

Field identification marks. (1) The fruit bodies large and coarse; (2) the gills are close; (3) the stalk originates deep in the soil and at the base one usually finds rudimentary fruit bodies which become arrested in their development; (4) it is typically found growing in clusters; (5) the cap is dingy white when young and fresh.

Observations. The small spores (for the genus) are an important microscopic character. The caps often push up through dirt such as when fruiting occurs at the edge of a road, and much dirt remains on the cap giving it a generally unattractive aspect.

Edibility. Edible. It is generally ignored by collectors because of its unattractive appearance.

When and where to find it. Common during the fall season along the Pacific Coast inland to northern Idaho and western Montana. At least one of its tree associates is Douglas fir.

Microscopic characters. **Spores** 10–13 (16) x 4.5–8.0 μ, narrowly elliptic in face view, subfusiform in profile, grayish

Gomphidius maculatus About natural size

3 *Gomphidius oregonensis* About one-half natural size

brown in KOH, yellow brown in Melzer's. **Basidia** 4-spored.
Pleurocystidia 80–120 x 8–14 (15) μ, cylindric to narrowly
clavate, with adhering material that is vinaceous in KOH (use
fresh material). **Caulocystidia** in fascicles near stipe-apex,
more variable in shape than the pleurocystidia. **Pileus
cuticle** of gelatinous appressed hyphae up to 2 μ in diameter.
Clamps absent.

Field identification marks.   (1) The subdistant, thick decurrent gills; (2) the slime veil readily observed on young fruit bodies; (3) the blackish spore deposit; (4) the slimy cap colored dingy purple brown to somewhat reddish cinnamon.

59 *Gomphidius glutinosus*   *About two-thirds natural s*

60 *Chroogomphus tomentosus*   *About two-thirds natural s*

Observations. A similar species with a ± rose red or pink cap is *G. subroseus.* The latter is common under Douglas fir. Both have a bright yellow stalk base and a slime veil. In Idaho I have found sterile specimens of *G. glutinosus* in which the gills remained white.

Edibility. Both *G. subroseus* and *G. glutinosus* are edible. Wipe the slime from the cap when you collect the fruit bodies.

When and where to find it. In our western area it is not uncommon in the spruce-fir zone of the mountain forests, and it is not uncommon to find it in quantity in spruce plantations. It fruits during the fall season.

Microscopic characters. **Spores** 15–20 x 4–7 μ, smoke brown under microscope, somewhat inequilateral in profile. **Pleurocystidia** 125–150 x 10–16 μ, prominent, subcylindric.

Chroogomphus

Chroogomphus tomentosus 60

Field identification marks. (1) A dry fibrillose dull yellow cap; (2) distant, thick yellow gills; (3) a dry stalk with a thin fibrillose veil; (4) the dark-colored spore deposit.

Observations. A second species with a grayish shade to the cap occurs with the above. It is *C. leptocystis.* The two often are found together. Both are edible.

Edibility. Edible.

When and where to find it. Both species are found under hemlock and Douglas fir, scattered but often abundant, especially in the Priest Lake district of Idaho, the Olympic National Park, and in the remains of the coastal conifer forests during the fall season.

Microscopic characters. **Spores** 15–21 x 6–7.5 μ, somewhat inequilateral in profile. **Pleurocystidia** 90–200 x 14–18 μ, subcylindric, walls thickened.

Chroogomphus rutilus 61

Field identification marks. (1) Caps ochraceous becoming olive brown to gray, and in age ± vinaceous; (2) cap surface sticky to the touch when fresh; (3) flesh of cap and stalk distinctly ochraceous; (4) gills ochraceous and thickish at least when young; (5) the spore deposit is dark gray to fuscous.

Observations. The fruit bodies illustrated are of the medium sized form, but very slender populations will also be encountered. *C. vinicolor* is readily distinguished by its thickwalled cystidia, those of *C. rutilus* being thin-walled.

Edibility. Edible and recommended by many. As far as known all species of *Chroogomphus* are edible.

When and where to find it. Scattered to gregarious under pine, especially on pine covered sand dunes along the coast

61 *Chroogomphus rutilus*   *About natural siz*

in November. It is to be expected throughout the pine areas
of the West.

Microscopic characters. **Spores** 14–22 x 6–7.5 μ, elliptic in
face view, subfusoid in profile. **Basidia** 4-spored. **Cystidia**
82–178 x 13–22 μ, narrowly fusiform. **Hyphae** of cap trama
at least in part amyloid. **Pileus cuticle** an ixocutis. **Clamps**
present.

Hygrophoraceae

Hygrophorus

The waxy gills and white or nearly white spore deposit
are the major features of this genus. The gills have
sharp edges in contrast to *Cantharellus*.

Key to Species

1.   KOH on the ornamentation at apex of stalk quickly
stains it orange; both a slime veil and fibrillose veil
lacking . . . . (p. 103) *H. pudorinus*

1.   Not as above 2

   2.   Cap gray to black; veil absent 3

   2.   Cap some other color 4

3.   Cap sticky; gills thin; odor of almonds present . . . . .
. . . . (p. 105) *H. agathosmus*

3.   Cap somewhat sticky but soon dry; gills thickish; odor
not distinctive . . . . (p. 104) *H. camarophyllus*

Hygrophorus pudorinus   62

Field identification marks.   (1) The cap is pale buff to pinkish or pinkish buff; (2) the markings over the apical region of the stalk quickly become orange with KOH and are reddish as dried; (3) the odor is slightly fragrant; (4) no veil is present.

Observations.   There is considerable variation in the color of the cap; it varies from whitish in one form to gray in another.

Edibility.   Rated highly for the European variants. Its quality relative to North American variants has not been adequately reported but it should be tested — some are bitter when cooked.

When and where to find it.   Under spruce or fir, late summer and fall, abundant in some coastal areas but generally rather unpredictable.

Microscopic characters.   **Spores** 6.5–9.5 x 4–5.5 μ, smooth, ellipsoid, inamyloid. **Basidia** 2- and (mostly) 4-spored. **Pleurocystidia** and **cheilocystidia** none. **Cuticle** of pileus an ixotrichoderm (a gelatinous turf) of hyphae 2–3 μ wide. **Clamps** present.

62 *Hygrophorus pudorinus*   *About two-thirds natural size*

Field identification marks. (1) The cap is blackish to dark, grayish brown (fuscous) and only slightly sticky when moist; (2) gills whitish to cinereous, distant and adnate to decurrent; (3) stalk dry and more or less colored like the cap (whitish if in deep moss) and equal or nearly so, very fragile; (4) absence of a veil of any kind.

Observations. The stalk is so tender one is likely to break it when collecting the species. *H. calophyllus,* which also occurs along the Pacific Coast, has colors somewhat like those of *H. camarophyllus,* but the cap is more slimy and the gills have a pinkish flush when older. It is most abundant along the northern coast of California. *H. marzuolus* is a species of the "snowbank flora" and rather similar in appearance to *H. camarophyllus.*

Edibility. This species is rated rather highly in Europe. It is a late-season fungus and worth some experimentation, but observe the usual precautions since the American populations do not check character for character with the descriptions of European material. It very soon becomes riddled by insect larvae. Finally, do not confuse it with *Clitocybe nebularis.*

When and where to find it. Solitary to gregarious in the spruce-fir zone west of the Cascade Divide, but not limited to this zone. I have seen it most abundantly in the Cascades around Mount Hood and Mount Rainier in late October or November.

Microscopic characters. **Spores** 7–9 x 4–5 μ, smooth, ellipsoid. **Basidia** 4-spored. **Cystidia** none. **Gill trama** divergent. **Cuticle** of pileus of gelatinous hyphae soon becoming collapsed and often difficult to demonstrate after wet weather. **Clamps** present.

63 *Hygrophorus camarophyllus*   *About two-thirds natural size*

Field identification marks.   (1) The distinct odor of bitter almonds (peach-pits when crushed); (2) the lack of either a fibrillose or slime veil; (3) the gray, sticky cap; (4) the lack of any ornamentation on the stalk; (5) lack of distinctive color changes on injury.

Observations.   *H. pustulatus* and *H. tephroleucus* are smaller and at maturity have ornamentation as gray spots or fibrils on the stalk. *H. calophyllus* has a pinkish reflection to the gills and lacks the odor of *H. agathosmus*.

Edibility.   Edible, but I have no information on its quality.

When and where to find it.   Usually around spruce trees in great abundance after heavy fall rains late in the season.

Microscopic characters.   **Spores** (7) 8–10.5 x 4.5–5.5 μ, ellipsoid, smooth, inamyloid. **Basidia** 4-spored. **Pleurocystidia** and **cheilocystidia** none. **Hyphae** of gill trama 5–10 μ broad, divergent. **Cuticle** of pileus a turf of gelatinous hyphae 1.5–4 μ wide. **Clamps** present.

64 *Hygrophorus agathosmus*   *Slightly less than natural size*

Hygrophorus speciosus   65

Field identification marks.   (1) The brilliant red to orange cap (it fades to yellow in aging); (2) a yellow slime veil leaving stains on the stalk; (3) occurrence under or near larch; (4) the gills are yellowish on the edges at least by maturity.

Observations. *H. aureus* of Europe, according to Fries, grows under hardwoods so this name is ruled out of consideration for the North American collections.

Edibility. Edible, but in 1970 I heard reports of a poisonous red, slimy *Hygrophorus* growing in central Oregon. Anyone *not* well acquainted with *Hygrophori* would do better to let others do the testing of edibility.

When and where to find it. Abundant in the fall where there is western larch. It has been found, however, in New Mexico where there is no larch.

Microscopic characters. **Spores** 8–10 x 4.5–6 μ, ellipsoid, smooth, inamyloid. **Basidia** 4-spored or 2-spored. **Pleurocystidia** and **cheilocystidia** none. **Hyphae** of gill trama 5–11 μ wide, divergent. **Cuticle** of pileus a slime layer 100–200 (300) μ thick, the hyphae 2–5 μ wide, more or less in a turf, colorless in KOH. **Clamps** present.

---

66 Hygrophorus gliocyclus

Field identification marks. (1) The cap is slimy and pallid with a yellowish disc; (2) the gills extend downward on the stalk and are close, whitish, and not spotting; (3) the stalk is slimy from the remains of an outer veil, and has an evanescent gelatinous ring; (4) the robust fruit body; (5) the association with spruce (?) and pine.

Observations. The context is very tender and soon worm-eaten. *H. flavodiscus* has the cap turning orange buff in drying, has smaller spores and gills flushed pinkish at first. *H. glutinosus* has red brown dots on the stipe when dried and, as found to date, occurs in deciduous woods.

Edibility. Edible and choice, at least so I have been informed by a number of collectors in northern Idaho where the species is quite abundant.

When and where to find it. Characteristically found in stands of mature lodge-pole pine late in the season (October–December). Rather abundant in northern Idaho during wet seasons.

Microscopic characters. **Spores** 8–10 (11) x 4.5–6 μ, ellipsoid, smooth, inamyloid. **Basidia** 4-spored. **Pleurocystidia** and **cheilocystidia** absent. **Hyphae** of gill trama 3–7 μ wide, divergent. **Cuticle** of pileus a zone 200–900 μ thick originally representing a turf but collapsing and in age the hyphae appressed, the hyphae 3–5 μ wide. **Clamps** present.

---

67 Hygrophorus saxatilis

Field identification marks. (1) The ochraceous watery spots usually in a zone near the margin of the cap; (2) lack of both a fibrillose and a slime veil; (3) odor reminding one of dried peaches; (4) the pinkish cinnamon to ochraceous salmon gills; (5) the stalk having a thinly appressed-fibrillose covering which is not a veil.

65 *Hygrophorus speciosus*   *Slightly less than natural size*

66 *Hygrophorus gliocyclus*   *About natural size*

Observations. *H. karstenii* is close but at least according to some authors is not sticky. *H. cremicolor* has gills approaching those of *H. saxatilis* in color but has an interwoven gill trama.

Edibility. Apparently not tested.

When and where to find it. Scattered under conifers on rocky soil of steep hillsides in particular but occurring generally in the mountain conifer forests of the region late in October or November. Common at times.

Microscopic characters. **Spores** 7–9.5 x 4–5 (6) μ, subellipsoid, smooth, inamyloid. **Basidia** 2- and 4-spored. **Pleurocystidia** and **cheilocystidia** none. **Hyphae** of gill trama 4–8 μ wide, divergent. **Cuticle** of pileus of appressed hyphae 2–3 μ wide, only slightly gelatinous, colorless. **Clamps** present.

67 *Hygrophorus saxatilis*   *Slightly less than natural size*

68 *Hygrophorus bakerensis*   *Slightly less than natural size*

Field identification marks.   (1) The lack of a slime veil; (2) the cap is some shade of crust brown over the center and paler toward the pallid margin; (3) the cap is sticky; (4) the odor is of crushed peach pits; (5) the gills are close, thin, and do not discolor when bruised; (6) when fresh, the gills and apex of the stalk typically are beaded with drops of a hyaline liquid.

Observations.   The thin gills may cause some to search in other genera for this species but the divergent gill trama clearly indicates *Hygrophorus.* The species has been observed repeatedly and the gills found to be bluntly attached to the stalk more often than extending down it. In drying the gills darken appreciably.

Edibility.   Not recommended. I have been informed by a number of collectors that the species is edible, but many of them thought it was a *Clitocybe* or a *Tricholoma!*

When and where to find it.   Common in the fall under conifers up to elevations of 4000 feet in the Cascades and coast ranges, and in the forested areas of the Rocky Mountains, especially in Idaho. It may occur scattered, gregarious, or clustered. The true firs appear to be among its associates.

Microscopic characters.   **Spores** 7–9 (10) x 4.5–5 (6) μ, smooth, ellipsoid. **Cystidia** none. **Gill trama** of divergent hyphae. **Pileus cuticle** a trichoderm of gelatinous hyphae the layer 100–230 μ thick. **Clamps** present.

Tricholomataceae

See key to families for a characterization. This is the largest and most complex family of the gilled fungi.

Key to Genera

1. Gills frequently forked, close and well formed . . . .(p. 122) *Hygrophoropsis*

1. Gills not forking except near the stalk, or only scattered forking occurring 2

   2. Gills yellow to orange, becoming olivaceous in the known western species; growing in clusters; gills phosphorescent when in good growing condition (see *Tricholomopsis* also) . . . .(p. 121) *Omphalotus*

   2. Not as above 3

3. Growing on wood; fruit body tough; gills with serrate edges at maturity . . . .(p. 123) *Lentinus*

3. Not with all three of above features .4

4. Stalk eccentric or absent 5
4. Stalk centrally attached to underside of cap (rarely ↖ eccentric) 7
5. Fruit body rubbery (not easily broken), with a gelatinized context or a gelatinized layer of tissue in or on the cap ....(p. 126) *Panellus*
5. Not as above 6
   6. Gills bright yellow to orange; spore deposit pink ....(p. 127) *Phyllotopsis*
   6. Gills usually whitish; spore deposit white, pale yellow, or lilac gray ....(p. 124) *Pleurotus*
7. Much mycelium (mold) present around the base of the stalk and permeating the duff; taste of context often bitter to disagreeable; spores amyloid; context of cap dry and firm ....(p. 128) *Leucopaxillus*
7. Not as above 8
   8. Cap surface dry and ± granulose; gills attached to the stalk ....(p. 141) *Cystoderma*
   8. Not as above 9
9. Veil (fibrillose or membranous) leaving a ring on the stalk when it breaks ....(p. 142) *Armillaria*
9. Veil absent to poorly developed 10
  10. Typically on wood; gills yellow; stalk present and central to eccentric ....(p. 136) *Tricholomopsis*
  10. Not as above 11
11. On soil; spore deposit pink; odor and taste strongly farinaceous; gills white at first ....see (p. 171) *Clitopilus*
11. Not as above 12
  12. Gills emarginate or adnexed; stalk 5–15(30) mm thick, typically growing from ground or humus ....(p. 130) *Tricholoma*
  12. Not as above 13
13. Fruit body pliant, drying out and reviving readily; stalk 0.5–3 mm thick ....(p. 149) *Marasmius*
13. Fruit body fragile or at least not readily reviving 14
  14. Gills strongly decurrent; cap often depressed over the disc; stalk 0.5–2(3) mm thick ....(p. 157) *Omphalina*
  14. Not as above 15
15. Gills ascending and adnate to stalked or attached with a tooth; cap cone-shaped; stalk 1–2.5 mm thick ....(p. 154) *Mycena*
15. Not as above 16
  16. Gills decurrent to broadly adnate; cap often vase-shaped in age; stalk over 3.5 mm thick at maturity ....(p. 111) *Clitocybe*
  16. Not as above 17
17. Gills adnate; cap ± convex and the margin inrolled at first; stalk 0.5–10 mm thick. Carminophilous granules

absent from basidia . . . .(p. 152) *Collybia*

17.   Fruit body with pallid to gray or blackish colors but basidia showing carminophilous granules (p. 147) *Lyophyllum**

*Note: *Lyophyllum* as a genus cannot be distinguished in the field from *Collybia* and *Clitocybe*. The collector, at this point, should consult the photographs and write ups of the two species of *Lyophyllum* included here.

Clitocybe

See key to genera for a characterization. *Clitocybe* is a large and complex genus in the West, in a sense competing with *Cortinarius* in its diversity.

Key to Species

1.   With a fleshy mass of fungous tissue at the base of the stalk or close to it (probably a parasitized *Helvella*) . . . .(p. 113) *C. sclerotoidea*

1.   Not as above 2

   2.   Gills green to bluish green . . . .(p. 114) *C. odora* var. *pacifica*

   2.   Not as above 3

3.   Caps 10–20 cm broad, pinkish tan to buff when fresh; gills whitish 4

3.   Not as above 5

   4.   Spores not amyloid . . . .(p. 114) *C. maxima*

   4.   Spores amyloid . . .see (p. 129) *Leucopaxillus septentrionalis*

5.   Caps dull bay red; context vinaceous in KOH; typically in clusters . . . . .see (p. 121) *Omphalotus olivascens*

5.   Not as above 6

   6.   Cap whitish or white; growing in dense masses along roads or on banks; cap often irregular . . . .(p. 117) *C. dilitata*

   6.   Cap gray, blackish, or dingy yellow brown 7

7.   Cap dingy yellow brown, with minute darker squamules near the margin; on conifer logs and stumps; spores amyloid . . . .(p. 117) *C. ectypoides*

7.   Cap gray, yellow brown to black (not as above) 8

   8.   Cap blackish and scurfy; gills white; on hardwood . . . .(p. 118) *C. atrialba*

   8.   Not as above 9

9.   Caps dingy yellow brown near margin, and the margin often ribbed; on or near rotting hardwood . . . .(p. 118) *C. avellaneialba*

9.   Not as above 10

   10. Cap, gills, and stalk violaceous when young; spore deposit vinaceous buff; gills typically adnate . .(p. 112) *C. nuda*

   10. Cap grayish over disc; gills white; spore deposit pale yellow . . . .(p. 121) *C. nebularis*

Field identification marks.   (1) The cap, gills, and stalk are violaceous when young; (2) there is never any sign of a veil (check young fruit bodies); (3) the spore print is pale vinaceous buff.

Observations.   This fungus has been generally known under the name *Tricholoma nudum* and *Lepista nuda*. The color is rather striking when the fruit bodies first develop but practically all the blue fades out leaving the fruit body with a dingy, pinkish brown tint. In the faded stage it is easily confused with other species of *Clitocybe* in the section *Verruculosae.* If the spore deposit has a strong reddish tone beware of the genus *Entoloma*. Its spores are angular and some of the species are poisonous.

Edibility.   Edible and popular.

When and where to find it.   It is a "trash inhabitor" and fares well in old leaf piles, compost piles, on humus in the woods, and around sawdust piles. It fruits from late summer on into the winter.

Microscopic characters.   **Spores** vinaceous buff (pale pinkish tan) in a deposit, 5.5–8 x 3.5–5 μ, ellipsoid, verruculose. **Cystidia** none. **Pileus cuticle** only slightly refractive in KOH (not truly gelatinous). **Clamps** present.

69 *Clitocybe nuda*   *About one-half natural size*

Field identification marks. (1) The fleshy mass of tissue at the base of the stalk (according to Trappe probably representing a *Helvella lacunosa*); (2) cap matted fibrillose and dingy pallid to grayish buff or dingy tan, at times with watery spots; (3) gills subdistant and colored much like the cap but finally gray ("drab"); (4) stalk surface with heavily matted tomentum (a covering of soft hairs).

Observations. In some respects this species appears related to *Clitocybe inornata* but is at once distinguished by the sclerotoid (fleshy) base.

Edibility. Not recommended.

When and where to find it. Typically in areas where *Helvella lacunosa* is known to fruit. It has been found frequently in the coastal area in Washington, Oregon, and northern California, often late in the fall. Presumably it is parasitic on the *Helvella*. To my knowledge it is known only from the area indicated.

Microscopic characters. **Spores** 8–10 (11) x 3–4 μ, subfusiform, smooth, hyaline. **Basidia** 4-spored. **Cystidia** none. **Cuticle** of pileus of hyphae 2–6 μ wide, the cells short, walls brownish in KOH but not incrusted. **Clamps** present.

70 *Clitocybe sclerotoidea*   *Slightly less than natural size*

71   Clitocybe odora var. pacifica

Field identification marks.   (1) The strong odor of anise (fragrant); (2) the dull green cap, stalk, and gills; (3) the ± adnate gills; (4) the yellowish color of the spore deposit is also an important feature.

Observations.   There are a number of variants around *C. odora* var. *odora* of which the above is the one most readily identified in the field in the western area because of its striking color.

Edibility.   Edible. Not recommended as a dish by itself because of the rather strong taste.

When and where to find it.   Scattered to gregarious under conifers during the fall season, generally distributed throughout the western area and often abundant. The most abundant fruiting I have ever seen was back of the old village dump at Coolin, Idaho.

Microscopic characters.   **Spores** 6.5–8 (9) x 4–5 μ, ellipsoid, smooth, hyaline, inamyloid, "pale pinkish buff" in deposit. **Basidia** 4-spored. **Pleuro-** and **cheilocystidia** none. **Pileus cuticle** a thin layer of hyphae 2–4 μ wide, cylindric and rather loosely interwoven. **Clamps** present.

72   Clitocybe maxima

Field identification marks.   (1) Caps 10–20 cm broad and shallowly funnel-shaped; (2) color when young a pinkish tan; (3) caps do not change color in fading; (4) odor and taste somewhat disagreeable; (5) stalk 1.5–3 cm at apex.

Observations.   It resembles *C. gibba* so closely that one is at times inclined to regard it as a variety of that species. It may be confused with species of *Leucopaxillus* by some, but the latter species have amyloid spores and fleshier caps. Also, most *Leucopaxillus* species have copious white mycelium permeating the surrounding duff.

Edibility.   Listed as edible in Europe. I have no data on its quality.

When and where to find it.   Solitary, gregarious or in small clusters in woods of conifers as well as hardwoods in our western area. It appears in the fall during warm wet seasons. Its distribution in the West remains to be ascertained. The best material I have seen came from near Lake Mills in the Olympic Mountains of Washington.

Microscopic characters.   **Spores** 7–9 (10) x 4–6 μ, smooth, in face view elliptic to pear-shaped, in profile almost dropshaped, inamyloid. **Basidia** 4-spored. **Cystidia** not differentiated. **Hyphae** of the pileus cuticle 2.5–5 μ wide, cylindric, appressed. **Clamps** present.

1 *Clitocybe odora* var. *pacifica*   *About natural size*

2 *Clitocybe maxima*   *About one-half natural size*

73 *Clitocybe dilitata*   *Slightly less than natural siz*

74 *Clitocybe ectypoides*   *Slightly less than natural siz*

Field identification marks. (1) The whitish cap with an ir-regular margin and often some zones present near the margin; (2) gregarious or growing in clusters on waste land, particu-larly on shoulders of dirt roads and on banks; (3) the some-what sour to disagreeable taste; (4) the gills typically remain-ing more or less adnate.

Observations. The white spore deposit and variously ir-regular cap distinguish it from *C. cerrusata.* There is hardly a collector who will not find this species in the western area. It can usually be spotted from one's auto as he drives along the road.

Edibility. Not recommended. Poisonous species occur in this group of *Clitocybes.*

When and where to find it. It fruits from late summer to winter throughout the area. See above for habitat data.

Microscopic characters. **Spores** 4.5–6 (6.5) x 3–3.5 μ, white in deposit, ellipsoid, smooth, inamyloid. **Basidia** 4-spored. **Cystidia** none. **Cuticle** of pileus of appressed hyphae 2–4 μ broad, cylindric. **Clamps** present.

Field identification marks. (1) The dingy tan cap with minute dark colored punctate scales at maturity; (2) habitat on decaying logs of conifers — especially hemlock; (3) the crowded, narrow, buff-colored gills.

Observations. Although this species was described from the state of New York, it is widely distributed and quite com-mon in the Pacific Northwest. The amyloid spores are an additional important feature which can be checked from a spore deposit.

Edibility. Not recommended.

When and where to find it. Solitary to gregarious or ces-pitose on decaying conifer wood — especially hemlock, dur-ing the fall season in the Pacific Northwest.

Microscopic characters. **Spores** 7–10 x 4–5 μ, ellipsoid; in deposit pale yellow. **Basidia** usually 4-spored. **Pleuro-** and **cheilocystidia** 2–5 μ wide, filamentose, scarce, projecting beyond the hymenium, up to 50 μ long. **Cuticle** of pileus of brown appressed hyphae 3–7 μ wide, apices often ascending in fascicles. **Clamps** present.

75 Clitocybe atrialba

Field identification marks. (1) The stalk is covered with dark particles of scurf; (2) the gills are distant to subdistant; (3) the cap is dark blackish brown and becomes furfuraceous; (4) it grows on wood of broad-leaved trees such as alder, maple, and cottonwood.

Observations. Add to the above that the spores are 9–11 x 7–9 μ and amyloid and an accurate identification can be made. In Singer's system the species is placed in *Clitocybula,* a genus including species removed from *Collybia* and *Clitocybe.*

Edibility. I have no records of its being tested.

When and where to find it. Scattered to clustered on buried logs or on wood above ground, as indicated above. It fruits rather early in the fall throughout most of the area west of the crest of the Cascades. Hardly ever does it occur in large numbers, but it can be collected repeatedly almost any season. It is endemic to the Northwest.

Microscopic characters. **Spores** 9–11 x 7–9 μ, broadly elliptic or (6) 7–9 (10) μ and globose, smooth, hyaline, amyloid. **Basidia** 2-spored or 1-spored. **Pleurocystidia** none. **Cheilocystidia** 43–60 x 9–14 μ, occasionally septate near base. **Pileus cuticle** with a decumbent turf of end cells with dark pigmentation. **Clamps** present.

76 Clitocybe avellaneialba

Field identification marks. (1) The habitat is on or close to rotten wood, especially of alder; (2) the cap is blackish brown gradually becoming paler to drab; (3) the gills are dull white; (4) the stalk is concolorous with the cap at first.

Observations. This species has much the aspect of *C. clavipes,* but its broad subfusoid spores and habitat on wood distinguish it at once. *C. clavipes* occurs under conifers, pine in particular.

Edibility. I have no data on it and experimentation is not recommended.

When and where to find it. Scattered, gregarious or subcespitose on rotting wood or on humus close to rotting wood under alder often with scattered conifers in the stand; common in the Pacific Northwest after the fall rains.

Microscopic characters. **Spores** 8–10 (11) x 4–5.5 μ, broadly fusoid, hyaline, smooth, inamyloid. **Basidia** 4-spored. **Pleurocystidia** and **cheilocystidia** not differentiated. **Pileus** with a thin cuticle of dark brown slightly gelatinized hyphae (in KOH), the pigment dissolved in the cell sap. **Clamps** regularly present.

75 *Clitocybe atrialba*   *About natural size*

76 *Clitocybe avellaneialba*   *About two-thirds natural size*

77 *Clitocybe nebularis*   *About one-half natural siz*

78 *Omphalotus olivascens*   *About one-half natural siz*

Field identification marks. (1) The large stature and gray to nearly pallid dry cap; (2) the adnate to short-decurrent crowded gills; (3) the peculiar and disagreeable odor and taste; (4) pale yellow spore deposit.

Observations. The surface is often bumpy due to an infection by a parasitic fungus, possibly a species of Mycogone. C. nebularis is also infected by another agaric, Volvariella surrecta.

Edibility. Not edible.

When and where to find it. This species often fruits in large numbers in the cedar-hemlock-alder areas throughout the Pacific Northwest during the fall season. It is to be regarded as a common species.

Microscopic characters. **Spores** yellowish in deposit, 6–8 x 3.5–4.5 μ, ellipsoid, smooth, inamyloid. **Basidia** usually 4-spored. **Cystidia** none differentiated on the gills. **Pileus** with a cuticle of hyphae 2–4 μ wide, rather thick. **Clamps** present.

Omphalotus

Field identification marks. (1) Caps dull bay red to the paler margin and becoming dull olivaceous in fading; (2) gills subdistant, broad, decurrent; (3) gills rather soon becoming dingy olivaceous with honey yellow edges; (4) stalk bright olive in the basal region, darker olivaceous upward.

Observations. Although the material was in good condition it looked "old." The fruit bodies are much redder dried than when fresh. Most other fungi in the woods at the time were just in the first stages of fruiting — Boletus satanus, B. regius, B. flaviporus, B. puniceus, and species of Cortinarius. In addition, with KOH the color on the fresh context was dark vinaceous.

Edibility. Not tested for obvious reasons. The species is closely related to Omphalotus olearia and O. illudens. O. illudens is to be expected in the oak areas of southern Oregon and California. Both are poisonous.

When and where to find it. Clustered on soil or road banks during the late fall, possibly coming from buried wood in the manner of Armillaria mellea.

Microscopic characters. **Spores** 5–6.5 x 4–5.5 μ, ellipsoid to subglobose, inamyloid, with yellow content in KOH. **Basidia** 4-spored. **Pleuro-** and **cheilocystidia** none found. Yellow droplets abundant, free in KOH mounts, and present in the hymenial layer, the latter yellow also.

*O. olivascens Bigelow and Thiers, in press.

79 Hygrophoropsis aurantiacus

Field identification marks. (1) The soft dry consistency of the cap; (2) the croceous to orange gills which are narrow, crowded, decurrent, and dichotomously forked; (3) the lack of a distinctive odor or taste.

Observations. The cap varies so much in color that it is not included as a field character of the species. The common form in the west has an orange to buff colored cap as shown in the photograph, and the gills in occasional collections are yellow rather than orange. The variation in the color of the cap is from pale orange to blackish brown. The species is generally considered to be a "dry weather" fungus. It is often abundant when most agarics are scarce.

Edibility. Reported as edible by some American authors, and listed as suspected by some in Europe.

When and where to find it. Solitary to gregarious on rich humus or around rotting wood or on burned areas in both conifer and hardwood forests in the fall. It is common in our western states.

Microscopic characters. **Spores** 6–8 x 4–4.5 μ, ellipsoid, smooth, rusty brown in Melzer's whitish to creamy in mass, walls thickened in age. **Basidia** 24–30 x 6–8 μ, 2-, 3-, or 4-spored. **Cystidia** not differentiated. **Cuticle** of pileus at first a turf of clavate to cystidioid end cells 7–16 μ wide, orange brown to yellow brown from dissolved pigment, the turf collapsing in age. **Clamps** present.

79 *Hygrophoropsis aurantiacus*   *About natural siz*

Lentinus kauffmanii 80

Field identification marks. (1) The eccentric stalk in most fruit bodies; (2) the rubbery-pliant consistency; (3) the vinaceous tinge of the cap when fresh and moist; (4) the acid then peppery taste; (5) the close, decurrent gills; (6) the habitat on conifer logs, especially Sitka spruce.

Observations. The gills often do not develop the characteristic serrations which feature the genus until full maturity. The species forms a brown pocket-rot in conifer logs.

Edibility. To my knowledge nothing is known of the edibility of this species. Ordinarily it is not found in large enough numbers to tempt one to try it.

When and where to find it. It is found solitary to scattered, rarely in small clusters along old conifer logs in the fall. According to my experience it favors Sitka spruce but is not limited to that species. The best fruitings I have found occurred when collecting for fleshy fungi generally was poor.

Microscopic characters. **Spores** 5–6 x 2–2.5 μ, smooth, inamyloid, oblong to elliptic in face view, in profile elliptic or with the ventral line (in optical section) straighter than the dorsal line. **Basidia** 4-spored. **Pleurocystidia** very abundant, 60–100 x 7–12 μ, hyaline in KOH, thin-walled, subcylindric to somewhat ventricose near base, rarely forked, sometimes with a secondary septum. **Cheilocystidia** very abundant, 60–125 x 5–9 μ, subequal to a slightly enlarged apex, smooth, hyaline. **Cuticle** of pileus a turf of more or less upright slender filaments 4–5 μ wide and becoming matted down to appear as a cuticle of appressed hyphae. **Clamps** regularly present.

Lentinus kauffmanii   *About natural size*

Key to Species

1. Growing on hardwood logs and stumps; cap gray to bluish gray . . . . (p. 124) *P. ostreatus*
1. Growing on conifer logs and stumps; cap white or weakly yellowish; gills often yellowish in age . . . .
. . . . (p. 125) *P. porrigens*

---

81   Pleurotus ostreatus
(Oyster Mushroom)

Field identification marks.   (1) The gray to bluish gray shelving caps; (2) the gills often fused at the stalk; (3) the fleshy (not cartilaginous) consistency; (4) habit of growing on stumps and logs of hardwoods such as elm and cottonwood.

Observations.   The gray variant, here illustrated, is often very robust in our western area. However, the picture was taken on an elm stump in Ann Arbor, Michigan. The spore deposit is grayish lilac in this variant. The "oyster mushroom complex" of variants in North America is still in need of more critical study.

Edibility.   Edible, but some variants in irrigated areas of our western states are reported to be unpalatable.

When and where to find it.   Occurring scattered on old logs to clustered-gregarious on old stumps, etc., often in large

81 *Pleurotus ostreatus*   *Less than one-fourth natural s*

rosettes or shapeless masses as shown in the accompanying photograph. It is most abundant late in the fall but may also be encountered in the spring. Common generally. It favors cottonwood and Lombardy poplar.

Microscopic characters. **Spores** 7–10 x 3.5–4 μ, grayish lilac in deposits, smooth, oblong in face view, inamyloid. **Basidia** 4-spored. **Cystidia** none. **Cuticle** of pileus of appressed hyphae 2–6 μ wide with thin smooth walls. **Clamps** present.

Pleurotus porrigens

82

(Angel's Wings)

Field identification marks. (1) Habitat on old conifer logs; (2) shining white to milk white fruit bodies 2–8 cm wide; (3) gills white at first but yellowing slightly by old age; (4) odor and/or taste not pronounced; (5) spore deposit white.

Edibility. Edible, and popular in the Pacific Northwest. It can often be collected in large quantities.

When and where to find it. In shelving masses on old conifer logs and stumps from late summer to late fall, common west of the Cascade Divide, and in northern Idaho.

Microscopic characters. **Spores** 6–7 x 5–6.5 μ, subglobose to globose, smooth, not amyloid. **Basidia** 4-spored, 30–37 x 7–9 μ. **Pleurocystidia** and **cheilocystidia** not differentiated. **Gill trama** regular-interwoven. **Pileus cuticle** not well differentiated (no gelatinous layer present). **Clamps** present.

82 *Pleurotus porrigens*   *About one-half natural size*

83   Panellus serotinus

Field identification marks.   (1) The dull green to bluish tints in the surface of the cap; (2) the gills and basal tubercle bright orange yellow when young; (3) the pliant consistency

83 *Panellus serotinus*   *About natural siz*

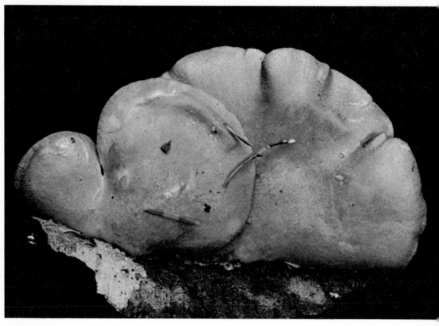

83 *Panellus serotinus*   *About natural si*

(caused by a gelatinous layer in the cap); (4) the mild odor and taste.

Observations. In North America two color "forms" occur, one olive green and the other duller and with bluish or violet tones. I have also found an occasional fruit body half olive and half dull violaceous. The reader is free to draw his own conclusions as to the taxonomic value of the difference.

Edibility. Edible and used by many people, but not choice. I have reports that it is bitterish when cooked.

When and where to find it. In my experience this species is the harbinger of the end of the mushroom season in late fall or winter depending on the region. It fruits abundantly on alder and other hardwood logs after prolonged cool wet weather in the western region.

Microscopic characters. **Spores** white in deposit, 4–5 (6) x 1.5 μ, smooth, cylindric in face view, curved (allantoid) in profile, inamyloid. **Basidia** 4-spored. **Pleurocystidia** abundant, 30–45 x 6–11 μ, fusoid-ventricose to subelliptic, thin-walled or some with thickened walls, with yellow or hyaline content. **Subhymenium** gelatinous. **Pileus** gelatinous in the cuticular region from upright narrow hyaline hyphae 4–5 μ wide; above this an epicutis of a layer of appressed hyphae 6–8 μ wide with end cells clavate to cystidioid and having the walls slightly thickened. **Clamps** present.

Phyllotopsis

Phyllotopsis nidulans   84

84 *Phyllotopsis nidulans*   *Slightly less than natural size*

Field identification marks. (1) Caps sessile, their surface fibrillose and dry; (2) bright orange to yellow; (3) the odor very disagreeable (with a sulphur component as in old cabbage); (4) gills close and bright orange to yellow; (5) spore deposit bright pink.

Observations. This is one of the easily recognized species of gilled fungi. Various authors give different spore sizes for it, a situation which indicates a need for further critical study.

Edibility. Apparently not poisonous but certainly not desirable either. Indications are, however, that it possesses some compounds of interest to chemists.

When and where to find it. In our western region it is to be expected during late summer or fall on aspen and alder in that order, but it is also reported (rarely) on conifers. It is a common species in the Great Lakes area but its occurrence westward has not been critically mapped. When it fruits, as a rule large numbers of fruit bodies are produced many of which may not ever reach maturity .

Microscopic characters. **Spore deposit** bright pink. **Spores** 6–8 x 3.5–4.5 μ, allantoid in profile view, cylindric in face view, smooth, not amyloid. **Clamps** present.

Leucopaxillus

Key to Species

1. Cap dark to light reddish brown . . . . (p. 128) *L. amarus*
1. Cap grayish leather color . . . . (p. 129) *L. septentrionalis*

85   Leucopaxillus amarus

Field identification marks. (1) The dry, dark to light reddish brown cap; (2) the persistently very bitter taste; (3) the white mold (mycelium) permeating the duff around the base of the stalk; (4) the close white adnate gills often extending into ribs down the apex of the stalk; (5) the white spore print, and amyloid ornamented spores.

Observations. There are a number of color variants in the species as described here: f. *roseibrunneus* with the color of the cap caused by a pigment dissolved in the cell sap (but some incrusted pigment also present on the hyphal walls), is one, and f. *bicolor* with the cap paler than the above variant and with less incrusted pigment on the hyphae is another. Dissolved pigment, when present, is in the cells of the cuticular hyphae.

Edibility. Not edible because of the taste.

When and where to find it. The type variant of the species illustrated fruits in the Englemann spruce zone of the Rocky Mountains during wet showery summer weather. It is not common; *roseibrunneus* is the common form west of the Cascade Divide and can often be collected in quantity.

85 *Leucopaxillus amarus*   *About one-half natural size*

Microscopic characters. **Spore print** white. **Spores** 4.5–6 x 3.5–5 μ, subglobose, ornamented with strongly amyloid warts. **Basidia** 4-spored. **Pleurocystidia** none. **Cheilocystidia** very abundant, 25–38 (46) x (2) 3–8 μ, fusoid, clavate or fusoid-ventricose. **Hyphae** of the pileus cuticle with both dissolved pigment in the cells and copious incrusting pigment on the hyphal walls, some cuticular hyphae with thickened walls. **Clamps** present.

Leucopaxillus septentrionalis   86

Field identification marks. (1) The odor of freshly collected specimens is strong and nauseous; (2) the stalk stains dingy buckskin color from handling; (3) the cap generally is a dingy tan when mature; (4) the stalk is 2.5–6 cm thick; (5) the gills are adnate to slightly adnexed.

Observations. This is a massive species; only the "buttons" are shown in the photograph. It is most closely related to *L. giganteus* and like *L. giganteus* at times may have caps 30 cm or more broad. It differs in odor and taste as described above, has adnate (not decurrent) gills, and the cap does not become funnel-shaped.

Edibility. Not recommended because of the disagreeable taste and lack of reliable data on whether it is actually poisonous or not.

When and where to find it. On duff under conifers, clustered, late fall after heavy rains. Not common. The most material I have seen was found on the eastern slope of the Cascades at Bear Springs in the Mount Hood National Forest.

Microscopic characters. **Spores** 6.5–9 x 4–5 μ, subovoid to ellipsoid, smooth, amyloid. **Basidia** 4-spored. **Pleuro-** and

cheilocystidia none. **Hyphae** of pileus cuticle narrow (3–8 μ), somewhat interwoven and appressed. **Clamps** present.

86 *Leucopaxillus septentrionalis*   *About one-half natural size*

Tricholoma

The fleshy stalk, white spore deposit, sinuate to ad-nexed gills and typically the absence of clamp connections on the hyphae of the fruit bodies characterize this genus. It is much more homogeneous than *Clitocybe*.

Key to Species

Tricholoma flavovirens   87

Field identification marks. (1) The sticky yellow cap becoming pale brownish over the disc; (2) the gills pale lemon yellow; (3) the stalk yellowish; (4) the veil absent.

Observations. There are a number of species (or variants) close to *T. flavovirens*. One of these has white gills. *T. flavovirens* was formerly well known under the name *Tricholoma equestre*.

Edibility. Edible and considered choice — but remember *Tricholoma* is a dangerous genus (see the poisonous *T. pardinum*).

When and where to find it. This species is associated with pine, often 2-needle pines. Consequently in the West it is found chiefly with lodgepole. Along the coastal sand dunes this is an important edible species but cleaning off the sand can be a chore. The fruit bodies are often found buried in the sand. It fruits during the fall season.

Microscopic characters. **Spore deposit** white. **Spores** hyaline in KOH, 6–7.5 x 4–5 μ, ellipsoid, smooth, inamyloid. **Basidia** 4-spored. **Pleurocystidia** and **cheilocystidia** none. **Pileus** with a well differentiated pellicle of narrow (2.5–4.5 μ), yellow, gelatinous hyphae, the pigment dissolved in the cell sap.

87 *Tricholoma flavovirens*   *Slightly less than natural size*

88   Tricholoma populinum

Field identification marks.   (1) It fruits in large masses (2 to 3 feet in diameter or in large fairy rings); (2) the cap is sticky and dark dull reddish cinnamon over the disc, the margin off-white; (3) the odor and taste are farinaceous; (4) it grows associated with cottonwood; (5) the gills stain vinaceous brown in age.

Observations.   I often speak and write about mushrooms fruiting in dense masses, and at times I have detected the attitude that such accounts were "fish stories." For this reason, a part of a large fairy ring is shown in the accompanying photo. Since the group to which this species belongs is a very complex group, collectors are advised to pay particular attention to the association with cottonwood.

Edibility.   Edible, popular in the Pacific Northwest — but be *sure* to collect it under cottonwood.

When and where to find it.   Late in the fall in fairy rings or arcs, cespitose-gregarious in the vicinity of cottonwood trees. One can usually fill a bushel basket from one fairy ring.

Microscopic characters.   **Spores** 5–5.5 x 3.5–4 μ, ellipsoid, smooth, inamyloid.

89   Tricholoma saponaceum

Field identification marks.   (1) The moist cap, and often the stalk sparsely scurfy; (2) the cap dominantly olive-colored; (3) the base of the stalk with pinkish orange tints in places when old; (4) the context staining yellowish where bruised.

Observations.   The variant of this species illustrated and described here is one of four which occur in our western area and, to judge from the literature, the species is equally complex in Europe. We have a second variant with a grayish brown cap and farinaceous taste which occurs in the alpine fir zone in the summer. All have the pink to orange stains at the base of the stalk in some fruit bodies of a collection, lack a veil, and have a moist smooth cap.

Edibility.   Not recommended. One should not eat any *Tricholoma* unless he is positive it is correctly identified. On this basis very few species of *Tricholoma* from our western area should be eaten.

When and where to find it.   Solitary to gregarious under spruce-fir stands in the mountain forests and forested sand dunes in the summer and fall, often abundant, but variants of this species may also turn up in the oak forests of southern Oregon and in California.

Microscopic characters.   **Spores** 5–5.5 x 3–3.5 μ, ovoid to ellipsoid, smooth, inamyloid. **Basidia** 4-spored. **Pleurocystidia** none. **Cheilocystidia** 10–18 x 2.5–3 μ or 15–20 x 6–7 μ, filamentose to clavate, subgelatinous as revived in KOH. **Pileus** with a pellicle of narrow interwoven subgelatinous hyphae (in KOH). **Clamps** present and readily demonstrated.

8 *Tricholoma populinum*   *About one-fourth natural size*

39 *Tricholoma saponaceum*   *About one-half natural size*

90 *Tricholoma pardinum*   *About one-half natural si*

91 *Tricholoma vaccinum*   *Slightly less than natural siz*

Field identification marks. (1) The gray scaly to fibrillose cap; (2) the robust stature; (3) the white unpolished stalk; (4) the distinctly farinaceous taste; (5) the whitish gills only rarely developing grayish stains on the edges; (6) a veil is lacking.

Observations. There is a large group of gray to fuscous species of *Tricholoma* having a dry fibrillose to squamulose cap. *T. pardinum* is one of the most robust of these; others such as *T. virgatum* have a tardily acrid taste.

Edibility. Poisonous. Under no circumstances should the amateur experiment with this group of fuscous to gray dry fibrillose species.

When and where to find it. Solitary to gregarious, during the fall season, often in large numbers in mixed, hardwood-conifer forests, and under pure stands of conifers, especially *Abies,* common during some years.

Microscopic characters. **Spores** 8–9 x 5–6 μ, ellipsoid, smooth, inamyloid. **Basidia** 4-spored. **Pleurocystidia** none. **Cheilocystidia** rare and embedded, clavate, up to 15 μ broad, hyaline, smooth. **Pileus** with a cuticle of radial hyphae with slightly thickened walls nearly hyaline in KOH. **Clamps** absent.

Field identification marks. (1) The caps are densely covered with cinnamon rufous scales; (2) the gills are pallid but soon stain vinaceous brown; (3) the stalk stains vinaceous tan to reddish when handled; (4) the odor and taste are not distinctive.

Observations. In stature this species may more or less resemble a large *Inocybe,* but the white spore deposit immediately distinguishes it. It is closest to *T. imbricatum,* but has a more scaly cap and more slender stature.

Edibility. Not recommended.

When and where to find it. Scattered to gregarious under conifers generally throughout the western area during late summer and fall, often quite abundant.

Microscopic characters. **Spores** white in deposit, 6.5–7.5 x 4.5–5 μ, broadly ellipsoid, smooth, inamyloid. **Basidia** 4-spored. **Pleurocystidia** and **cheilocystidia** none seen. **Pileus cuticle** of radial hyphae grouped in fascicles with darker orange brown walls in KOH but the filaments not appreciably thinner than those of the context. **Clamps** absent.

Field identification marks.   (1) The caps are dingy dull cinnamon to dull vinaceous brown; (2) cap appressed-fibrillose becoming slightly scaly near the margin; (3) taste slightly mealy.

Observations.   This species has a more robust stature than *T. vaccinum,* a much less scaly cap, and the taste is somewhat mealy. Both show the same type of color change and the pigment pattern is much the same.

Edibility.   Not recommended. It is listed as edible by some authors, but in a dangerous genus such as *Tricholoma* it is better not to take chances.

When and where to find it.   Solitary to scattered in conifer and mixed conifer-hardwood forests generally throughout the western area during the fall season.

Microscopic characters.   **Spores** 5.5–7 x 4–4.5 μ, ellipsoid, smooth, inamyloid. **Basidia** 4-spored. **Pleuro-** and **cheilocystidia** not differentiated. **Pileus cuticle** of radially arranged hyphae with dull reddish brown walls and 5–8 μ wide.

92 *Tricholoma imbricatum*   *Slightly less than natural si*

Tricholomopsis

This genus was erected to include species taken from *Tricholoma, Clitocybe* and *Collybia* of the Friesian system. The habitat is typically on decaying wood, cheilocystidia are present and often very large, clamp connections are present and the cap is typically fibrillose to some degree. Yellow figures prominently in the pigmentation of the fruit body in nearly all species.

Key to Species

1. Cap with purple red fibrils over the disc . . . (p. 137) *T. rutilans*
1. Not as above 2
    2. Fruit body yellow but with gray to blackish fibrils and / or scales over the disc . . . . (p. 138) *T. decora*
    2. Not as above 3
3. Cheilocystidia 40–200 x 3–5 μ, filamentose; gill edges darker yellow than the faces . . . . (p. 138) *T. flavissima*
3. Cheilocystidia 46–75 x 12–18 μ, clavate to fusoid-ventricose; gill edges concolorous with faces . . . . . .
    . . . . (p. 140) *T. sulfureoides*

Tricholomopsis rutilans   93

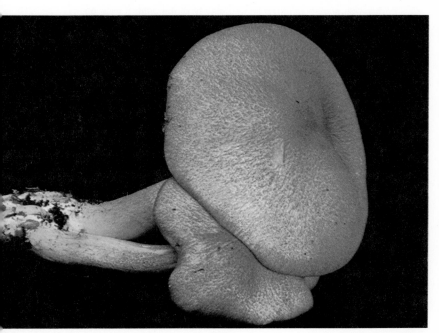

Tricholomopsis rutilans   *About natural size*

Field identification marks. (1) The cap is covered by purplish red fibrils breaking up into fibrillose scales; (2) a strong yellow undertone is evident by maturity; (3) the gills are yellow; (4) the stalk is colored about like the cap.

Observations. This is a well-known species included here to show variations. The epicuticular fibrils on the cap vary to purplish red to vinaceous red to almost ferruginous.

Edibility. Edible, but of inferior quality.

When and where to find it. Scattered to cespitose on conifer logs, common in the fall throughout the western area.

Microscopic characters. **Spores** 5–6 x 3.5–4 μ, white in deposit, broadly ellipsoid, smooth, inamyloid. **Basidia** 4-spored. **Pleurocystidia** 36–52 (60) x 6–9 μ, clavate to subcylindric, more refractive than basidioles. **Cheilocystidia** very abundant, clavate, fusoid-ventricose or mucronate to fusoid (40) 50–80 (110) x 12–20 (30) μ, dingy reddish brown in KOH, thin-walled. **Pileus cuticle** composed of enlarged and elongate cells with red content. **Clamps** present.

94   Tricholomopsis decora

Field identification marks. (1) The cap has a yellow ground color and on this, especially over the disc, occur fuscous (blackish) fibrils or scales; (2) the yellow more or less adnate gills; (3) the yellowish stalk; (4) growing on the wood of conifers.

Observations. There are a number of yellow species in this genus, in fact it is likely that the presence of this pigment can be regarded as a generic characteristic. The grayish to fuscous scales or fibrils limited mainly to the central area on the cap are the key character of this species. Usually only a few fruit bodies are found at a time.

Edibility. Edible.

When and where to find it. Solitary to gregarious on rotting conifer logs of various species in late summer and fall; common in the western area but seldom abundant.

Microscopic characters. **Spore deposit** white. **Spores** 6–7.5 x 4.5–5 μ, subellipsoid, smooth, hyaline in KOH, inamyloid. **Basidia** 2- and 4-spored. **Pleurocystidia** rare, 34–42 x 5–8 μ, subcylindric, projecting only slightly beyond the basidia. **Cheilocystidia** 36–62 x 9–20 μ, clavate, saccate or fusoid-ventricose, yellowish to orange in KOH. **Scales** of pileus of fascicles of hyphae with slightly incrusted walls. **Clamps** present.

95   Tricholomopsis flavissima

Field identification marks. (1) The bright yellow color over all; (2) the slightly peppery taste; (3) the gills having brighter yellow edges than do the faces; (4) the rather slender stature.

Observations. *T. fulvescens* is close to *T. flavissima* but the cheilocystidia distinguish them: They are clavate for *T. fulvescens.*

Edibility. Not known.

When and where to find it. Scattered to subcespitose on wood of conifers in the fall in Oregon, Washington, and Idaho. Uncommon.

4 *Tricholomopsis decora*   *Slightly less than natural size*

5 *Tricholomopsis flavissima*   *About natural size*

Microscopic characters. **Spores** globose to subglobose, (6) 7–9 x 6–8 μ, smooth, inamyloid. **Basidia** 4-spored. **Pleurocystidia** none. **Cheilocystidia** very abundant, 40–200 x 3–5 μ, septate, filamentous, subgelatinous, yellowish to hyaline when fresh (the pigment intracellular). **Epicutis** of pileus of greatly elongated hyphal cells (not inflated), the walls thin or thickened, smooth. **Clamps** present.

Field identification marks.   (1) The pale yellow cap varie-
gated with paler areas and streaks; (2) the yellow, broad,
adnexed to adnate gills; (3) the yellow stalk; (4) the presence
of a thin veil in immature fruit bodies.

Observations.   *T. bella* and *T. thompsoniana* are very closely
related but lack a veil. The variants around these species
have as yet not been completely documented for the western
region. Differences in the presence or absence of cystidia
and in spore size and shape indicate the presence in our
flora of additional taxa.

Edibility.   I have no reliable data on it.

When and where to find it.   Solitary to gregarious on conifer logs, often hemlock; found in northern Idaho and probably throughout the hemlock zone in the western area during
the fall rainy season.

Microscopic characters.   **Spores** 5.5–6.5 x 4.5–5 μ, smooth,
elliptic in face view, obscurely kidney-shaped in profile,
inamyloid. **Basidia** 4-spored. **Pleurocystidia** scattered, projecting up to 12 μ, 35–47 x 6–9 μ, clavate to fusoid-ventricose,
hyaline. **Cheilocystidia** clavate to fusoid-ventricose, 46–75 x
12–18 μ, wall yellow in KOH. **Cuticle** of pileus of radial
hyphae, the end cells cylindric. **Clamps** present.

96 *Tricholomopsis sulfureoides*   *About natural si*

Cystoderma

Key to Species

1. Ring well-formed and flaring; odor not distinctive when fresh . . . .(p. 141) *C. fallax*
1. Ring in an annular zone and additional veil remnants on stalk below it; odor strong of freshly husked green corn . . . .(p. 142) *C. amianthinum* f. *rugosoreticulatum*

Cystoderma fallax   97

Field identification marks.   (1) The evenly colored subferruginous to amber brown cap; (2) the well-formed flaring ring; (3) the lack of a distinctive odor; (4) the veil remnants on the stalk extending down from the ring to the base and of the same color and texture as the cap surface.

Observations.   The small, amyloid spores and lack of cystidia aid further in making an accurate identification.

Edibility.   Not tested as far as I am aware.

When and where to find it.   Scattered to gregarious in moss under conifers; often abundant in the fall season throughout the western area.

Microscopic characters.   **Spores** 3.5–5.5 x 2.8–3.6 (4) μ, broadly ellipsoid, amyloid, smooth. **Basidia** 4-spored. **Pleuro-** and **cheilocystidia** none. **Cuticle** of pileus of chains of inflated to globose cells with dark cinnamon walls as revived in KOH. **Clamps** present.

Cystoderma fallax   *About natural size*

98   Cystoderma amianthinum f. rugosoreticulatum

Field identification marks.   (1) The strong odor of freshly husked green corn; (2) the pronounced radial wrinkling of the cap; (3) the yellow to yellow brown cap (not having distinct reddish tones); (4) the slender stalk lacking a persistent ring (of the type shown for *C. fallax),* but with the veil material scattered along it in zones and patches; (5) the habitat in mossy conifer forests.

Observations.   If one has dilute potash solution (KOH–2.5 percent) with him and places a drop on the cap of this species the spot becomes dark rusty brown.

Edibility.   I have no information on the edibility of the variants of this species. There is hardly enough "body" to the fruit bodies to attract the mycophagist.

When and where to find it.   Common in mossy areas, especially conifer forests at upper intermediate elevations in the mountains after heavy fall rains.

Microscopic characters.   **Spores** 5–6 (7.5) x 3 μ, amyloid, smooth, white in deposit, narrowly ellipsoid. **Basidia** 4-spored. **Pleuro-** and **cheilocystidia** not differentiated. **Cuticle** of pileus of globose to variously shaped inflated cells 12–50 x 10–30 μ, revived in KOH having tawny brown walls, walls smooth and only slightly, if at all, thickened. **Clamps** present.

Armillaria (Including *Armillariella)*

Gills are attached to the stalk, the stalk has a partial veil the remains of which form a ring. The spore deposit is white to yellowish; the stalk is fleshy; the cap and stalk are not readily separable; and the cuticle of the cap is of appressed fibrils. These are the characteristics of this genus.

Key to Species

1.   Growing in clusters on or very near old logs or stumps or close to trees; cap honey brown to blackish brown ....(p. 144) *A. mellea*

1.   Growing on humus solitary to gregarious 2

2.   Cap white at first, tinged cinnamon when older ....(p. 146) *A. ponderosa*

2.   Cap bright yellow to dull yellow on margin and center brownish 3

3.   Cap bright yellow and conspicuously scaly ....(p. 143) *A. luteovirens*

3.   Cap dull yellow and center often smoky brown, not conspicuously scaly ....(p. 144) *A. albolanaripes*

Cystoderma amianthinum f. *rugosoreticulatum*   *About natural size*

Armillaria luteovirens   99

Field identification marks.   (1) The bright yellow fruit bodies
with conspicuous innate yellow scales on the cap; (2) the
scaly stalk below the annular zone (the scales are yellow);
(3) the mild odor and taste.

Armillaria luteovirens   *Slightly less than natural size*

Observations.   The weakly amyloid spores are an important additional feature and add to the evidence that *A. albolanaripes* and it are closely related.

Edibility.   Not tested as far as I am aware — at least not in North America.

When and where to find it.   Solitary to gregarious under mixed conifers and hardwoods during the fall in Oregon. It is apparently rare in the western area and not known elsewhere in North America.

Microscopic characters.   **Spores** 6.5–8 x 4–5 μ, ellipsoid to oblong, smooth, weakly amyloid. **Basidia** 4-spored. **Pleuro-** and **cheilocystidia** none seen. **Pileus** with an epicutis breaking up into fascicles of more or less parallel hyphae, the fascicles forming the scales of the pileus. **Clamps** present.

---

100   Armillaria albolanaripes

Field identification marks.   (1) Yellow cap with brownish center and more or less scaly over marginal area; (2) yellow gills which are depressed-adnate; (3) conspicuous remains of veil on stalk below the ring.

Observations.   *A. luteovirens* is close to *A. albolanaripes* but the former has a more conspicuously scaly cap, typically bright yellow overall. In the southern Rocky Mountains there occurs a variant with a grayish cap. All of these have amyloid spores though in *A. albolanaripes* the reaction is very weak and not always evident until the specimens have aged for a time in the herbarium.

Edibility.   I know of no cases of poisoning from it, but any one trying it should observe the usual precautions.

When and where to find it.   Solitary to gregarious on humus under conifers and alder. It fruits in the summer in the Rocky Mountains (Salmon River country of Idaho), and in the fall in the Coastal area. It is rare to absent during some seasons.

Microscopic characters.   **Spores** 6–8 x 4–4.5 μ, smooth, ellipsoid, hyaline inamyloid at first, weakly amyloid in the herbarium. **Basidia** 4-spored. **Pleuro-** and **cheilocystidia** none. **Pileus** with a cuticle of appressed hyphae 5–8 μ wide, not gelatinous. **Clamps** occasionally present.

---

101   Armillaria mellea (Honey Mushroom)

Field identification marks.   (1) In clusters near or on wood of any kind — conifer or hardwood; (2) a somewhat cottony to membranous ring above the middle of the stalk; (3) gills staining (discoloring) dingy brown in age; (4) spore deposit white, at least in the western collections I have tested.

Observations.   The basal third of the stalk is often rusty brown. The species is difficult to describe because it is so variable in appearance, but if one remembers that it is the only clustered white-spored mushroom with a ring growing on wood, one is not likely to go astray.

Armillaria albolanaripes

Slightly less than natural size

1 Armillaria mellea

About one-half natural size

Edibility. Edible and choice — and very common.

When and where to find it. It fruits in the fall around living or dead trees, around stumps, on logs, etc. It is often a root parasite and a very unwelcome sight in the area where nuts are grown, such as in southern Oregon.

Microscopic characters. **Spore deposit** white. **Spores** 7–9 x 5–6 μ, smooth, nonamyloid. **Cheilocystidia** fusoid-ventricose to clavate, 28–40 x 7–10 μ. **Pleurocystidia** none. **Pellicle** of pileus somewhat gelatinized. **Clamps** absent to rare (check base of basidium).

102 Armillaria ponderosa
(Pine Mushroom)

Field identification marks. (1) The overall white color but with the cap becoming cinnamon tan; (2) the spicy odor; (3) the crowded adnexed gills; (4) annulate stalk; (5) large size and fleshy consistency; (6) typically growing near pine trees.

Observations. One can hardly write a guide to western mushrooms without including this species since it is collected commercially in the area. It is a most popular species with our Oriental friends, a situation no doubt arising from the fact that in Japan a closely related fungus, the Matsutake, is also very popular. In fact I have heard them refer to our species as the "white Matsutake."

Edibility. Edible and choice, but the simple procedure of sautéing it in butter is not the recommended method of preparation. Consult the recipes in the books listed in the chapter on edibility (p. 12).

When and where to find it. It fruits in the fall season on the pine covered sand dunes of the coastal region. The fungus can be found throughout the pine areas of the region.

Microscopic characters. **Spores** 5.5–7 x 4.5–5.5 μ, smooth, nonamyloid, ellipsoid to subglobose. **Hyphae** of the pileus cuticle 5–10 μ wide, not gelatinized. **Clamps** absent.

102 *Armillaria ponderosa*   *About one-half natural s*

Lyophyllum

This genus is composed mainly of species from *Clito-cybe, Collybia,* and *Tricholoma.* In the broad sense it includes all of these species which give a positive (purple) reaction on granules in the basidium when tested with ferricacetocarmine stain. This test will not be practical for most who use this book. However, the amateur should pay attention to the following: (1) Do the gills stain (discolor) to black when bruised (they may discolor to yellow or blue before going to black)? If this change occurs, the species is a *Lyophyllum* in the strict sense of the word. (2) If the colors of the cap are in the dull white to pale, medium, or dark gray or brownish gray to fuscous or black color range, the specimen is very likely to be a *Lyophyllum* if the stalk is more than 3 mm thick and the cap is hygrophanous (changes color in fading).

Key to Species

1. Clustered; cap 4–12 cm broad; gills close to crowded; often around piles of organic debris . . . . (p. 147) *L. decastes*
1. Not clustered and caps smaller; near melting snow-banks; odor and taste mild; stalk 10–16 mm thick, not rooting . . . . (p. 148) *L. montanum*

Lyophyllum decastes   103

Lyophyllum decastes   *About one-half natural size*

Field identification marks. (1) The fruit bodies are densely clustered; (2) the color of the cap is dark to pale tan or dark gray at first, but not blackish; (3) the gills are broadly adnate to somewhat decurrent; (4) the odor and taste are mild.

Observations. The name given above is used here to include a substantial number of variants as Singer has outlined them. The group needs further study.

Edibility. Edible and choice, according to collectors in both the Great Lakes area and the western states. It is likely that all the variants are being collected and eaten.

When and where to find it. The species fruits in large clusters often gregarious on waste grassy land, vacant lots during wet weather in the late summer and fall. One can often harvest it by the bushelful.

Microscopic characters. **Spores** globose, 4–6 μ, smooth, inamyloid. **Basidia** 4-spored. **Clamps** present.

---

104 Lyophyllum montanum

Field identification marks. (1) The habit of fruiting at the edge of melting snowbanks or in areas where the snow has just recently melted; (2) the gray gills and cap; (3) the lack of a distinctive odor and taste; (4) the gray, canescent stalk.

Observations. The species of this genus are almost impossible to distinguish without using a microscope, but for *L. montanum* the habit of fruiting near melting snowbanks is a most important character. The fruit bodies gradually change greatly in appearance as they age because of being in such a cool location. Deterioration is slow to set in, and hence various degrees of weathering are to be expected.

104 *Lyophyllum montanum*   *About one-half natural s*

Edibility. Not recommended. It has not been adequately tested to my knowledge, and, as pointed out, one can easily get over-aged specimens.

When and where to find it. *L. montanum* is one of the classical species of what has become known as the "snowbank mushroom flora" of our western mountains. This flora consists of an assemblage of quite unrelated species adapted to the peculiar features of habitat and climate — cold seepage moisture, intense light during the day, frosty nights, and the ability to form mycorrhizae possibly with both *Abies lasiocarpa* (alpine fir) and *Picea engelmanni* (Engelmann spruce). This group fruits as the snowbanks recede upward through the forested zone. The mycelia obviously are distributed throughout the spruce-fir zone.

Microscopic characters. **Spores** 6.5–8 x 3.5–4 μ, oblong to narrowly ellipsoid, inamyloid, white in deposit. **Basidia** 4-spored. **Pleurocystidia** and **cheilocystidia** absent. **Cuticle** of pileus a poorly defined subgelatinous layer of appressed hyphae 2–5 μ wide. **Clamps** present.

Marasmius

Fruit bodies with very thin pliant flesh, reviving when moistened; stalk narrow mostly under 5 mm thick; spore deposit white.

Key to Species

1. Growing in circles in lawns and meadows; stalk 3–7 mm thick . . . . (p. 149) *M. oreades*
1. Growing on wood or debris in the forest 2
    2. Stalk short and mostly eccentric; occurring in small to extensive clusters or masses on sticks and small branches . . . . (p. 150) *M. magnisporus*
    2. Stalk elongate (3–8 cm), central; growing solitary to scattered; gills narrow and merely subdistant . . . . . .
        . . . . (p. 151) *M. umbilicatus*

Marasmius oreades (Fairy Ring)   105

Field identification marks. (1) No veil is present; (2) the gills are exposed during their entire development and at maturity are broad, thickish, and spaced rather far apart; (3) the stalk is slender and tough; (4) the spore deposit is white; (5) the cap is pale buff to reddish tan and when water-soaked often marked by irregular lines.

Observations. This fungus is a well-known pest to lawn lovers in the northwest. It produces a zone of dead grass

105 *Marasmius oreades*

Slightly less than natural size

and a zone of stimulated growth — thus causing an interruption of the even green color in the lawns that some house owners try to maintain.

Edibility. By many rated as one of the best edible fungi — but discard the stalks. Also, beware of a small sordid-appearing *Clitocybe (C. dealbata)* with close decurrent gills which often grows with *M. oreades.* It causes profuse perspiration if eaten.

When and where to find it. As stated, both species grow in grassy places, lawns in particular, during the summer and fall.

Microscopic characters. **Spores** 7–9 x 4–5.5 μ, smooth, inamyloid, in profile somewhat inequilateral. **Basidia** 4-spored, reddish brown in Melzer's. **Pleurocystidia** none. **Cheilocystidia** 26–34 x 3–5 μ, hyaline in KOH, red brown in Melzer's. **Pilear trama** homogeneous beneath a cuticle of clavate cells arranged in a hymeniform layer, the cells dark red brown in Melzer's. **Clamps** present.

106   Marasmius magnisporus

Field identification marks. (1) It is entirely white when young and in age often discolored reddish; (2) the gills are distant, adnate to decurrent and at maturity moderately broad; (3) the stalk is white at first but darkens from the base upward; (4) characteristically fruiting on small branches of alder, canes of *Rubus,* dead branches of vine-maple, salal, etc.

Observations. In some works this fungus is treated under the name *M. candidus.*

Marasmius magnisporus   *About natural size*

Edibility.   Of no consequence as an edible fungus.

When and where to find it.   Common, and in dense masses on decaying small branches of hardwood trees, dead canes of *Rubus* and dead canes of *Gaultheria shallon* (salal) during the fall season in the coastal region.

Microscopic characters.   **Spores** 10–13 x 5–6 μ, white in deposit, narrowly drop-shaped, smooth, thin-walled, inamyloid. **Basidia** 4-spored. **Pleurocystidia** and **cheilocystidia** similar, 40–80 (140) x 7–10 μ, subcylindric or slightly ventricose near the base, hyaline, thin-walled. **Pileus trama** practically homogeneous; **pileocystidia** present, similar to pleurocystidia or with thick-walls, some setalike and 150–200 μ long.

Marasmius umbilicatus   107

Field identification marks.   (1) Cap 2–4 cm broad (large for a *Marasmius*), dull white except for very young stages which are pale buff; (2) surface more or less wrinkled; (3) gills whitish, narrow, and subdistant; (4) stalk 3–8 cm long and 1.5–3.5 mm thick, dull white at apex and finally dark brown below, white pruinose and unpolished; (5) odor and taste not distinctive.

Observations.   The cap is not umbilicate in most of the material I have collected. This species is one of the most conspicuous *Marasmiae* in the Pacific Northwest.

Edibility.   Not known — it has little to attract the mycophagist because of the very thin flesh and tough stalk.

When and where to find it.   During many seasons in the

Pacific Northwest one encounters 1–4 specimens at a time, but during other seasons, it occurs by the hundreds. In my experience it is a fall feature of areas in the conifer forests where thickets of Devil's Club are found.

Microscopic characters. **Spores** white in deposit, 9–12 x 3–3.5 μ, in face view narrowly subfusoid, pointed at apiculate end, smooth nonamyloid, thin-walled. **Basidia** 4-spored. **Pleuro-** and **cheilocystidia** 50–72 x 5–8 μ, acicular to sub-fusoid smooth, thin-walled, hyaline in KOH. **Cuticle** of pileus of interwoven hyphae furnished with short branches and projections (which cause the dull appearance of the surface). **Clamps** present.

107 *Marasmius umbilicatus* ·   *About natural si*

Collybia

Key to Species

1. Cap 3–7 cm broad; stalk 2.5–5(7) mm thick, lacking short lateral branches, surface velvety . . . .
    . . . .(p. 153) *C. cylindrospora*

1. Cap 3–10 mm broad; stalk ± 1–1.5 mm thick, with short lateral branches . . . .(p. 153) *C. racemosa*

Field identification marks. (1) The wide thin cap in relation to the narrow stalk; (2) the glabrous tan-colored cap; (3) the thin pliant flesh of the cap tending to revive when moistened.

Observations. This species, as for *Collybia confluens,* could just as well be placed in *Marasmius* — depending on your concept of the latter. The small narrow spores are a microscopic character of aid in recognizing it along with the features given above.

Edibility. To my knowledge not yet tested.

When and where to find it. Gregarious to subcespitose under alder in Oregon and Washington in particular during late summer and fall.

Microscopic characters. **Spores** 5–6 x 2.5 μ, narrowly ellipsoid, smooth, inamyloid. **Basidia** 4-spored. **Pleurocystidia** none. **Cheilocystidia** scattered, 18–30 x 2.5–3 μ, flexuous and filamentose. **Pileus cuticle** of appressed hyphae slightly narrower than those of the context.

108 *Collybia cylindrospora*   *About natural size*

Field identification marks. (1) The short branches covering the lower half or two thirds of the stalk; (2) the fuscous to dark gray disc of the cap (the margin is usually a paler gray); (3) habit of growing on decaying fruit bodies of agarics

109 *Collybia racemosa*   *About natural size*

(note that the host may be so far decayed as not to be recognizable as the remains of a mushroom); (4) the fruit body arising directly from a black body called a sclerotium which is part of the parasite, not the host.

Observations.   The sclerotium is not shown in the picture because none were found attached to the specimens photographed, most collectors will miss the sclerotium the first time they collect the species. The spores produced on the enlargement at the tips of the branches are known as conidia (asexual spores) and represent a means of spreading the species rapidly. Sometimes one will find stalks with many side branches but no caps at the top.

Edibility.   Of no consequence.

When and where to find it.   In clusters of 2–10 fruit bodies on decaying mushrooms during the summer, fall, and often later in the season after a heavy fruiting period for mushrooms is about over.

Microscopic characters.   **Spores** 4–5 x 2.5 μ, oblong, smooth, hyaline, inamyloid. **Basidia** 4-spored. **Pleurocystidia** and **cheilocystidia** none. **Pileus** lacking a differentiated cuticle. **Clamps** present.

Mycena

Key to Species

Field identification marks. (1) The bright coral pink cap; (2) the lack of a distinctive odor and taste; (3) the white to pinkish gills with pallid edges; (4) the naked stalk tinged coral pink at first.

Observations. A second brilliant coral pink to red species is *M. monticola,* but the cystidia are clavate and the apex covered with rodlike projections. *M. strobilinoides* has marginate gills and more scarlet to orange pigmentation. It is often very abundant under alpine fir in the Cascades and Olympics.

Edibility. Not of any consequence.

When and where to find it. Gregarious on moss in conifer forests and in *Sphagnum* bogs, not uncommon after heavy rains in the late summer and fall.

Microscopic characters. **Spores** 7–9 x 3–4 μ, smooth, ellipsoid, inamyloid. **Basidia** 4-spored. **Pleurocystidia** and **cheilocystidia** similar, 40–65 x 8–15 μ, fusoid-ventricose, neck slender, apex subacute to acute, hyaline smooth, thin-walled.

110 *Mycena amabilissima*   *About natural size*

Mycena aurantiomarginata 111

Field identification marks. (1) The dark brown cap with orange tints at the margin; (2) the orange margins of the gills (which slowly fade to orange yellow); (3) growing gregariously often in large numbers under spruce.

111 *Mycena aurantiomarginata*   *About natural size*

112 *Omphalina ericetorum*   *About natural size*

Observations. The color of the gill edges may become more general in vigorous specimens and spread to the faces. The pigment is located in the cell sap of the cystidia and hyphae; it is not incrusted on the walls.

Edibility. Not recommended. Some in this genus are poisonous; I know this from personal experience.

When and where to find it. Scattered to gregarious under conifers, especially Sitka spruce along the Pacific Coast. It fruits, often in large numbers, during the fall rainy season.

Microscopic characters. **Spores** 7–9 x 4–5 μ, smooth, ellipsoid, amyloid. **Basidia** 4-spored. **Pleurocystidia** and **cheilocystidia** abundant and similar, 28–36 x 7–12 μ, clavate to subcapitate, the apices sparsely or densely echinulate, cell sap bright orange. **Pileus trama** covered by a thin pellicle on which occur scattered pileocystidia similar to the cheilocystidia.

Omphalina

The incurved margin of the cap, the decurrent gills, and the cartilaginous stalk are the important characters. The spore deposit is white or yellowish, and the fruit bodies do not revive well when remoistened. It is distinguished from *Mycena* with some difficulty.

Key to Species

1. Cap olive over disc and lemon yellow near margin . . . . (p. 159) *O. wynniae*
1. Cap not as above 2
   2. Cap bittersweet orange (very bright) to orange buff or when faded yellow . . . . (p. 158) *O. luteicolor*
   2. Cap vinaceous brown, in age dingy yellowish pallid along the margin . . . . (p. 157) *O. ericetorum*

Omphalina ericetorum   **112**

Field identification marks. (1) The dull cinnamon to vinaceous brown color of the fresh cap; (2) the wide, distant gills; (3) the dull colored naked stalk; (4) the habitat on algal-covered or lichen-covered logs.

Observations. The name in current use is given above but some problems remain to be solved relative to our western variant. We have two variants, one has a yellow spore deposit, and the other a white deposit. This was verified from a dozen prints, and a critical comparison with white-spored variants shows only minor additional differences.

Edibility. Of no consequence.

When and where to find it. Solitary to scattered on old conifer logs covered by algae or lichens — such as those in light
shade in pastures, at edges of fields, old slashings, etc.
Common but never in great abundance, found in the fall.

Microscopic characters. **Spores** 7–9 (10) x 4–6 (7) μ (up to
12 x 8 μ on 1-spored basidia), broadly ellipsoid, smooth, inamyloid. **Cystidia** none. **Cuticle** of pileus a thin layer of appressed hyphae 2–5 μ wide. **Clamps** none.

113 Omphalina luteicolor

Field identification marks. (1) The combination of bitter-
sweet orange cap and orange buff gills; (2) the sparse fibrils
often causing the margin of the cap to be minutely fringed
at first (not shown in the photo); (3) the orange stalk; (4) the
habitat on old conifer logs or trunks.

Observations. The pigment in the cap apparently is water
soluble because the original bright orange pink soon becomes diluted to salmon buff to orange ochraceous. If one
wishes to get all the color stages, when you find it on a log,
look for some fruit bodies on the underside of the log or
where protected by loose bark.

Edibility. Not tested.

When and where to find it. Characteristically it is on old
butt logs left in the woods because of showing defect at the
time they were cut. It is common in the fall in the Northwest,
and frequently large numbers of fruit bodies are produced.

Microscopic characters. **Spores** 8–10 x 4–5 μ, smooth, ellipsoid, inamyloid. **Basidia** 4-spored. **Pleuro-** and **cheilocystidia**
not present. **Gill trama** of interwoven inamyloid hyphae.
Cuticle of pileus not sharply differentiated, of appressed
hyphae about like those of the context. **Clamps** absent.

113 *Omphalina luteicolor*   *About natural size*

Field identification marks.   (1) The cap is glabrous at first, lemon yellow near the margin, and with olive tones over the disc; (2) it has yellow, distant, long-decurrent gills; (3) the stalk is naked and yellowish; (4) the habitat is wet, partly decayed conifer wood.

Observations.   *O. ochroleucoides,* which also occurs on wood, is cream color in all parts. *O. ericetorum* also lacks the bright colors of *O. wynniae* and grows in association with the lichen *Botrydina vulgaris. O. wynniae* is very closely related to *O. luteicolor* but differs in lacking red tones.

Edibility.   Not worth considering.

When and where to find it.   Scattered to gregarious or in small clusters on rotten, water-soaked wood of conifers during the fall season.

Microscopic characters.   **Spores** 7–9 (10.5) x 4–5 (6) μ, elliptic or ovate to elliptic-oblong, smooth, not amyloid, color of deposit not known. **Basidia** 4-spored, 28–35 x 5.5–7.5 μ. **Cystidia** not differentiated. **Pileus cuticle** of loosely interwoven cystidial end cells, cylindric to subclavate, subventricose or mucronate, 5.5–18 μ wide. **Clamps** absent.

114 *Omphalina wynniae*   *About natural size*

Lepiotaceae

Key to Species

1. Spore deposit olive to green . . . .
. . . . (p. 162) *Chlorophyllum molybdites*
1. Spore deposit white or whitish 2
  2. Cap dry, fibrillose, white overall or disc grayish; stalk
    4–8 (10) mm thick . . . . (p. 160) *Leucoagaricus naucinus*
  2. Not as above 3
3. Cap large (5-15 cm) soon coarsely scaly; flesh slowly
    becoming brown where injured; stalk 1-2 cm thick . . .
    . . . . (p. 161) *Leucoagaricus rachodes*
3. Cap smaller, scales finer than in above; stalk slender
    (3-5 mm thick) and ± covered with floccose material
    up to where the veil breaks . . . . . . (p. 163) *Lepiota clypeolaria*

Leucoagaricus

---

115 Leucoagaricus naucinus

Field identification marks. (1) The cap is white or grayish
over the center, silky and dry to the touch; (2) the gills are
white but in age become grayish pink; (3) the gills are not
attached to the stalk and the latter is easily separated from
the cap; (4) the well formed ring on the stalk; (5) the base of

115 *Leucoagaricus naucinus*   *About one-half natural size*

the stalk is not enclosed in a cup or sheath (a volva) as in the white species related to *Amanita virosa*.

Observations. This species is easily confused with *Chlorophyllum molybdites* unless a spore deposit is obtained. The gills in *C. molybdites* may remain white a long time, and *L. naucinus* specimens with olive gills have been encountered. Hence the emphasis is on the spore deposit.

Edibility. Generally rated as a good edible species, but I still hesitate to recommend it because of the danger of confusing it with poisonous species.

When and where to find it. Scattered to gregarious on lawns and in grassy places, often abundant in warm rainy weather generally throughout the area.

Microscopic characters. **Spore deposit** white. **Spores** 7–9 x 5–6 μ, smooth, dextrinoid, with a germ pore. **Basidia** 4-spored. **Pleurocystidia** none. **Cheilocystidia** abundant, 25–40 x 7–12 μ, clavate, saccate, or fusoid-ventricose. **Gill trama** interwoven. **Pileus** lacking a distinct cuticle.

Leucoagaricus rachodes 116

Field identification marks. (1) Cap is large and coarsely scaly; (2) the stalk stains reddish then brown; (3) the ring is large and movable; (4) the spore print is white. It is very important to determine the color of the spore deposit — see *Chlorophyllum molybdites*.

Observations. In the Northwest where this species is frequently encountered, the green-spored *C. molybdites* is seldom if ever encountered. It may, however, turn out to be abundant some season in the future, just as has happened

16 *Leucoagaricus rachodes*   About one-half natural size

for the deadly *Amanita phalloides* in California. The mush-room hunter must be on guard against such occurrences at all times.

Edibility.   Edible and highly rated, as most collectors in the Northwest already know from personal experience.

When and where to find it.   Solitary to gregarious often under old spruce trees, on waste land, along roadsides during late summer and fall.

Microscopic characters.   **Spores** 8–10.5 x 5–6.5 μ, subel-lipsoid to somewhat ovoid, strongly pseudoamyloid but when first mounted with a distinct vinaceous tint, with a small apical lens-shaped plate. **Basidia** 22–34 x 7–9 μ, clavate, four-spored. **Pleurocystidia** none. **Cheilocystidia** abundant, vesiculose, clavate, or mucronate, 18–27 x 10–15 μ, brownish to hyaline in KOH. **Gill trama** loosely floccose. **Pileus cuticle** of a com-pact hymeniform layer of clavate to capitate cells 18–36 x 8–14 μ, smooth, with flexuous pedicels, the layer dull pale brown in KOH.

Chlorophyllum

117   Chlorophyllum molybdites

Field identification marks.   (1) The large whitish cap with tan to dark brown scales over the central area (or the layer forming a skull-cap at first); (2) the thick ring on the stalk and the pale olive to green color of the spore deposit; (3) the stalk separates cleanly and rather easily from the cap.

117 *Chlorophyllum molybdites*   *About one-half natural size*

Observations. The race which occurs in our Southwest has clamp connections on the hyphae of the fruit body and deserves further study. In spite of all the comments in the literature warning people about this species, poisonings by it still occur. This is because the gills often become green slowly. They are white, then olive yellowish and finally grayish green.

Edibility. To be regarded as poisonous. Species closely resembling it are definitely *not* recommended. In other words, if you have any doubts as to the identity of a collection at hand, do not eat it.

When and where to find it. In fairy rings in grassy areas or on waste land in the summer and fall. In our Southwest it is sometimes found during the rainy period in the summer. We need more data on its distribution in our western area.

Microscopic characters. **Spores** mostly 9–12 x 6.5–8 μ, smooth, dextrinoid, apex ± truncate. **Pleurocystidia** absent. **Cheilocystidia** present, 35–60 x 10–18 μ, clavate to fusoid-ventricose with obtuse apex, content hyaline. **Clamps** present.

Lepiota

Lepiota clypeolaria 118

Field identification marks. (1) The lacerate-scaly cap surface over the marginal half; (2) the smooth crust brown to tawny central area; (3) the dry consistency of the cap flesh; (4) the ragged zone of fibrils rather than a true membranous ring on the upper part of the stalk; (5) the ragged-floccose

118 *Lepiota clypeolaria*   *Slightly less than natural size*

material adhering along the stalk down from the ragged ring; (6) the whitish gills.

Observations. A number of color variations of this species occur in North America, but the common western variant is a robust rather highly colored (ochraceous to tawny) variant.

Edibility. Not recommended. *Lepiota* is a genus in which some distinctly poisonous species are found along with edible ones. It is best that one who does not know the genus refrain from testing them.

When and where to find it. Late summer and fall, often in second-growth conifer stands of pole-sized trees. It is rather general in its distribution in the Northwest, and often abundant.

Microscopic characters. **Spore deposit** white. **Spores** 13–18(20) x 4.5–6 μ, fusoid, dextrinoid. **Basidia** 4-spored. **Gill trama** of interwoven hyphae. **Cuticle** of pileus of ± upright hyphal elements with brown walls in KOH, the end cells 100–125 μ long and forming fascicles (the scales of the cap). **Clamps** present on some hyphae.

Amanitaceae

Gilled fungi with bilateral gill trama, free gills, white spore deposit, and with either an outer or an inner veil (usually both).

Amanita

This genus characteristically includes gilled fungi with gills free from the stalk, remains of an outer veil often remaining around the base of the stalk as a volva, usually with a ring representing remains of a partial veil, the stalk is central, the spores are typically white in deposit, and the gill trama is of divergent hyphae.

Key to Species

1. Volva membranous and persistent as a cup at base of stalk 2
1. Outer veil breaking up into particles, warts or a powder, remains at first present on cap and around base of stalk (but easily removed) 4
  2. Cap pure white . . . .(p. 170) *A. bisporigera*
  2. Cap yellowish to olive 3
3. Cap usually with a patch of membranous volval tissue over or near the disc; volva thick and rigid, with margin usually free from stalk and appearing double . . . . . .
  . . . .(p. 165) *A. calyptroderma*
3. Cap glabrous; volva thin and collapsing on stalk . . . .
  . . . .(p. 166) *A. phalloides*

Amanita calyptroderma   119

Field identification marks.   (1) The large size, thick outer veil
which leaves a skullcaplike patch on the cap; (2) the ochre
to ochre brown color of the cap; (3) thick rigid volva; (4) lack
of color changes where handled.

Observations.   This is perhaps the best known *Amanita* on
the Pacific Coast. The important point concerns people collecting puffballs for the table. The "eggs" of young puffballs
should be homogenous throughout, if in the edible stage. The
Amanita buttons show the outline of the mushroom when
sectioned longitudinally.

Edibility.   Apparently one of the best edible fungi known,
but I refuse to recommend it because it is an *Amanita*.

When and where to find it.   Scattered to gregarious under
conifers and hardwoods, along the Pacific Coast, during late
summer on into winter, most abundant in California.

Microscopic characters.   **Spore deposit** white. **Spores** (8)
9–11 x 5–6 μ, nonamyloid, hyaline in KOH, smooth ellipsoid.
Basidia 4-spored. **Pleurocystidia** none. **Cheilocystidia** clavate
to saccate-pedicellate, 42–64 x 8–16 μ, numerous. **Pileus
cuticle** a thick gelatinous pellicle, the hyphae 2–3 μ in diam.

119 *Amanita calyptroderma*   *About one-half natural size*

120   Amanita phalloides

Field identification marks.   (1) The cap varies from olive fuscous on the disc to paler olive or olive yellow over the marginal area; (2) the volva is membranous forming a conspicuous cup around the basal bulb; (3) the spores are amyloid; (4) an ample superior membranous ring is present; (5) the cap lacks veil remnants.

Observations.   The cap color may vary to dull yellow with a smoky disc, and rarely a piece of the volva may be left over the center. One should *beware* of any species with a well formed ring high up on the stalk and a "death cup" (volva) at the base. To get the latter be sure to dig out the mushroom carefully.

Edibility.   THE DEADLY POISONOUS SPECIES!

When and where to find it.   Our knowledge of this species in North America has followed a most interesting pattern. During the last part of the last century, and the first quarter of the present one, various species were reported under the name *A. phalloides.* One in particular was finally described as *A. brunnescens.* It, along with *A. citrina* and *A. porphyria,* have a less distinct volva than *A. phalloides* and the bulb in all three often splits perpendicularly. However in recent years the true *A. phalloides* appeared both in California and Oregon in the West and New Jersey and Pennsylvania in the East. As is usual with luxuriant fruitings of this fungus, cases of poisoning occurred, some of them fatal. The account presented here is based on American material, including the photograph by Mr. Donald Simons which is reproduced here. The dried specimens are the same, as far as I can determine, with material I collected in Belgium. It is to be expected during late summer to late fall.

120 *Amanita phalloides*   *About one-half natural size*

Microscopic characters. **Spore deposit** white. **Spores** 7–10 x 6–8.5 μ, globose to subglobose, smooth, thin-walled, amyloid, hyaline in KOH. **Basidia** 45–55 x 10–12 μ, 4-spored, clavate. **Pleurocystidia** absent. **Cheilocystidia** 24–31 x 5–12 μ, hyaline, clavate to saccate. **Gill trama** divergent from a central strand, inamyloid. **Pileus** with a gelatinized pellicle of narrow (2–3 μ) hyphae.

Amanita muscaria 121

Field identification marks. (1) The red to orange or yellow cap with particles of outer veil tissue scattered over it; (2) the outer layer of tissue over the lower part of the stalk typically becomes broken into zones or patches; (3) the outer veil is intergrown with the tissue of the bulb hence there is no free volva; (4) the ring is typically high up on the stalk.

Observations. A white form is sometimes encountered and difficulty may be encountered in distinguishing it from certain typically white species.

Edibility. Poisonous. There is a long history and much literature on this species, but my advice to the beginner is that he "blacklist" it. A cult has developed in connection with the hallucinogenic properties of this species.

When and where to find it. Scattered to gregarious under aspen and conifers, especially pine, rarely under oak; spring and fall, common throughout the West. In the Rocky Mountains it is often abundant in the summer.

Microscopic characters. **Spores** 9–11 x 7–8 μ, broadly ellipsoid, nonamyloid, with a large central oil drop. **Basidia** 4-spored, 44–52 x 7–9 μ. **Pleurocystidia** none. **Cheilocystidia** 38–62 x 8–12 μ, saccate to clavate, or some secondarily (?) septate and the cells ± enlarged. **Gill trama** of diverging hyphae from a central area. **Pellicle of pileus** gelatinized.

121 *Amanita muscaria*   *About one-half natural size*

122 Amanita pantherina

Field identification marks.   (1) A gray brown to dingy yellow or yellow brown cap; (2) creamy white warts or particles of outer veil tissue scattered over the cap surface (these often wear away by maturity); (3) a basal bulb with a collar around its apex (representing a volva intergrown with the tissue of the stalk).

Observations.   Until *A. phalloides* came into the picture prominently this species was considered the most poisonous species in the West. It is also one of the most variable, and there has been speculation as to whether it hybridizes with other species. Certainly strains exist which vary in their degree of toxicity and in the color of the cap.

When and where to find it.   Fruits during the spring (usually rarely), summer in the northern Rocky Mountains, and mostly in the fall. To be expected throughout the range of *A. muscaria*. It was particularly abundant in the Tacoma Prairies years ago.

Microscopic characters.   **Spores** 9–11 x 6.5–8 μ, smooth, broadly ellipsoid, not amyloid. **Pileus cuticle** a gelatinous pellicle, the hyphae 3–4 μ wide.

123 Amanita silvicola

Field identification marks. (1) Fruit body white overall (rarely ± discolored); (2) outer veil tissue leaving floccose remnants of it scattered variously over cap surface; flesh lacking a color change when cut or bruised; (3) stalk at first with a distinct bulb at base with its surface bearing floccose remains of outer veil (as for the cap); (4) ring on the stalk not well formed and soon vanishing; (5) absence of any distinctive odor.

Observations.   The stalk of this species is typically a short one, in contrast to many species in the genus.

Edibility.   I do not recommend any *Amanita,* even those I know to be edible.

When and where to find it.   Solitary to scattered in conifer and mixed forests, often along old roads, late summer and fall, not abundant, but frequently collected in the Pacific Northwest.

Microscopic characters.   **Spores** 9–12 x 5–6 μ, ellipsoid, slightly amyloid, smooth, white in deposits, thin-walled. **Basidia** 4-spored. **Pleurocystidia** none. **Cheilocystidia** (10) 15–30(40) x (9)12–20(30) μ, saccate, clavate, pear-shaped or ± globose-pedicellate, thin-walled, hyaline. **Gill trama** divergent. **Cuticle** of pileus a thin gelatinous pellicle. **Clamps** not found.

22 *Amanita pantherina*   *About one-half natural size*

23 *Amanita silvicola*   *About one-half natural size*

Field identification marks. This is one of the famous "destroying angel" group of *A. verna, A. virosa, A. phalloides,* and others. It has (1) a more slender and generally smaller fruit body than either *A. verna* or *A. virosa. A. phalloides* is distinct by its yellowish olive to olive cap. The critical character, as the species name indicates, is (2) that each basidium bears only two instead of four spores.

Edibility. DEADLY POISONOUS. It is more poisonous gram for gram of tissue than either *A. verna* or *A. virosa,* but of course there is considerable variation in chemical constituents within all three species. Good supportive treatment in time saves many lives. Thioctic acid as an antidote does not appear as effective as first indicated. The prognosis in this country for curing a patient is still uncertain.

When and where to find it. It is to be expected in the birch-aspen areas, especially in the Priest Lake district of Idaho, in the fall.

Microscopic characters. **Spores** are ± globose. **Basidia** 2-spored.

124 *Amanita bisporigera*   *About natural size*

Clitopilus prunulus   125

Field identification marks.   (1) The close, decurrent gills with a vinaceous tone at maturity; (2) the grayish, dry, unpolished cap; (3) the strong farinaceous odor of the crushed flesh; (4) the reddish spore deposit.

Observations.   I have tried to recognize two species in North America: (1) the above, and (2) *Clitopilus orcella* — the latter with a white, somewhat sticky cap as contrasted to the dark gray to cinereous colors of *C. prunulus,* but it must be admitted that intermediate collections such as the one illustrated here are frequent. In this collection the youngest caps were pale cinereous. The gray variant was not uncommon in the Sitka spruce forests along the Oregon coast during the season of 1970.

Edibility.   Edible and choice. The real danger in eating this species is that of getting a grayish capped species of *Entoloma* by mistake. *Entoloma* has angular spores and *C. prunulus* has longitudinally striate spores.

When and where to find it.   Solitary to gregarious on humus in the woods and often on sandy soil in late summer and fall; not uncommon but generally not abundant. I have seen it in large fruitings near Pacific City, Oregon.

Microscopic characters.   Spores 10–12 x 5–7 μ, subfusoid, with distinct longitudinal furrows, reddish in a deposit. **Basidia** 4-spored. **Cystidia** none. **Clamps** absent.

125 *Clitopilus prunulus*   *About natural size*

Volvariaceae

126   Volvariella surrecta

Field identification marks.   (1) Its occurrence on fruit bodies of other agarics, mostly on *Clitocybe nebularis;* (2) the white color overall when young; (3) the dry silky cap; (4) the reddish gills at maturity caused by the color of the spores (which are reddish in deposit).

Observations.   The only other agarics growing parasitically on other mushrooms and likely to be found in the region are species of *Nyctalis (Asterophora* in some works). Both *N. asterophora* and *N. parasitica* have rather thick gills — if indeed gills are present — whereas in *V. surrecta* the gills are thin, well-formed, and become reddish in age from the spores. *Psathyrella epimyces* grows on *Coprinus comatus.*

Edibility.   Many species of *Volvariella* are good (the "paddy-straw" mushroom of oriental commerce belongs to this genus). *V. surrecta,* however, is not found in sufficient quantity to be regarded as an esculent.

When and where to find it.   Found in the western area in the fall mostly on fruit bodies of *Clitocybe nebularis* (which is common). Obviously its time of appearance varies with that of the host.

Microscopic characters.   **Spores** 6–7.5 x 3.5–5 μ, ovoid, smooth, nearly hyaline under the microscope. **Basidia** 4-

126 *Volvariella surrecta*   *About natural size*

spored. **Pleurocystidia** 25–60 x 10–30 μ, fusoid-ventricose, some enlarged at the apex, varying to fusoid, clavate or ovoid. **Cheilocystidia** 25–60 x (6) 10–20 (40) μ, fusoid-ventricose, the neck short, varying to clavate or obovoid. **Pileus cuticle** of hyphae 5–20 μ wide, not gelatinous. **Clamps** apparently absent.

Cortinariaceae

This family is here delimited on a broader basis than in most of the current technical literature, and covers the rusty brown-spored gilled fungi with the exception of the Paxillaceae. It intergrades with the Strophariaceae.

Key to Genera

1. Veil absent; stalk deeply rooting . . . . . (p. 190) *Phaeocollybia*
1. Not as above 2
   2. Stalk 1–2.5 cm thick, furnished with a membranous ring; terrestrial and scattered to gregarious . . (p. 188) *Rozites*
   2. Not as above 3
3. Young fruit body possessing a cobwebby veil (cortina) and rarely an outer veil in addition; stalk typically over 5 mm thick; terrestrial . . . . (p. 174) *Cortinarius*
3. Not as above (but see *Inocybe* also) 4
   4. Gills white-crenulate on edges and beaded with droplets; stalk about 1 cm thick or more; cap sticky; veil absent (note that species with a veil are separated from *Cortinarius* on microscopic characters) . . . (p. 192) *Hebeloma*
   4. Not as above 5
5. Spore deposit bright orange to bright rusty fulvous; growing on wood or (rarely) on duff (if on duff, the veil absent in the one species included here) . . (p. 194) *Gymnopilus*
5. Spore deposit dull rusty brown to earth brown 6
   6. Typically growing on wood or burned ground near charred wood; stalk 3–20 mm thick and usually showing remains of a fibrillose to membranous veil variously disposed (see *Galerina* also) . . . . (p. 199) *Pholiota*
   6. Not as above 7
7. Stalk fleshy and usually over 3 mm thick; terrestrial; cap moist to dry but not sticky (some are slightly tacky when wet) . . . . (p. 194) *Inocybe*
7. Stalk 1–3 mm thick; on moss, wood, or humus . . . . . .
   . . . . (p. 197) *Galerina*

Cortinarius

This, the largest genus of gilled fungi in North America, has been neglected in all popular treatments of mushrooms simply because the North American species are not adequately known. The genus features rusty brown roughened (warty to wrinkled) spores, a cobweblike inner veil (the cortina), and the stalk and cap being confluent and hence not readily separable from each other. There are about 800 species in North America. The Pacific Northwest has an exceptionally rich flora.

Key to Species

11. Cap whitish when young but soon staining yellow to orange and cut context soon stained likewise . . . . (p. 182) *C. rubicundulus*

11. Not as above 12

   12. Growing in sand near pine; cap fibrillose and dull fulvous; gills becoming dull orange . . . . (p. 188) *C. aureifolius*

   12. Not as above 13

13. Cap rusty red when moist, paler when faded . . . . (p. 187) *C. californicus*

13. Cap buff to alutaceous, not fading and changing color as in above species . . . . (p. 187) *C. subaustralis*

Cortinarius vibratilis   127

Field identification marks.   (1) The fulvous to pale tan slimy cap; (2) the white, clavate, slimy stalk; (3) the bitter taste of the context and pellicle of the cap; (4) the generally small size; (5) whitish gills when young.

Observations.   This species is included here to illustrate one of two types of fruit body found in the subgenus *Myxacium* of *Cortinarius:* namely the one with clavate stalks.

Edibility.   Not recommended. The bitter taste should discourage most people.

7 *Cortinarius vibratilis*   *Slightly less than natural size*

When and where to find it.   Scattered to gregarious in mossy conifer forests, especially hemlock during the fall season in the West, but widely distributed and not limited to an association with hemlock.

Microscopic characters.   Spores 7–8 x 4.5–5 μ, warty-rugulose, ellipsoid, ochraceous-tawny in KOH. **Basidia** 4-spored. **Cystidia** none. **Pellicle** of pileus of narrow gelatinous hyphae 2–5 μ wide. **Clamps** present.

---

128   Cortinarius superbus

Field identification marks.   (1) The strong odor of freshly husked green corn; (2) the context staining brown; (3) the yellow gills; (4) fibrillose patches on the stalk darkening on their surfaces; (5) yellow cap when young.

Observations.   This is one of the most distinctive species of *Cortinarius* in the West. One may find the stalk sticky at the base from slime which drips off the cap, but this is not sufficient to cause one to search in subgenus *Myxacium* in trying to identify his collection.

Edibility.   Not known.

When and where to find it.   Solitary to gregarious in old-growth Douglas fir-hemlock forests in Idaho, Washington, and Oregon during the fall.

Microscopic characters.   Spores 11–13 x 5.5–7.5 μ, warty-rugulose, inequilateral in profile, dark rusty brown in KOH. **Basidia** 4-spored. **Cystidia** none (except for colored bodies in the hymenium). **Trama of pileus** purplish in KOH; **pellicle** a thick gelatinous layer. **Clamps** present.

---

129   Cortinarius pallidifolius

Field identification marks.   (1) The pallid gills when young; (2) tawny cap; (3) the slime sheath on the stalk brownish ochraceous and separating into zones or patches; (4) the clavate stalk; (5) a dense white fibrillose sheath beneath the slime layer.

Observations.   This is one of the common and distinctive species of the conifer forests of the Pacific Northwest and not easily confused with other species.

Edibility.   Not known.

When and where to find it.   Scattered to gregarious under fir *(Abies),* but its association with that genus possibly not rigid. It is a feature of the fall mushroom season in the mountains of the Northwest.

Microscopic characters.   Spores 9–12 (13) x 5–6.5 μ, warty-roughened, dark rusty brown in KOH, inequilateral in profile. **Basidia** 4-spored. **Pleuro-** and **cheilocystidia** none found. **Cuticle** of pileus a layer of yellowish (in KOH) gelatinous hyphae with clamps at the cross walls.

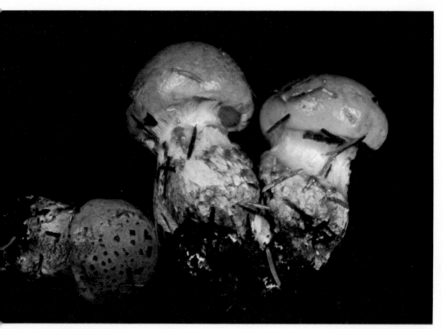

28 *Cortinarius superbus*   *About one-half natural size*

29 *Cortinarius pallidifolius*   *About two-thirds natural size*

130 *Cortinarius bigelowii*   *About natural siz*

131 *Cortinarius subfoetidus*   *About two-thirds natural siz*

Field identification marks. (1) The fruit bodies are produced underground or occasionally they break through; (2) thin but persistent veil breaking only in age; (3) pale alutaceous to yellowish cap; (4) grayish (avellaneous) gills when young; (5) stalk with a basal marginate bulb.

Observations. This species is an example of a peculiar type of ecological adaptation known for several endemic species of *Cortinarius* in our western region. It is not a "gastromycete" because the spores are discharged from the basidia (a spore print is easily obtained).

Edibility. Not known, and one should not experiment with *Cortinarius* since few species have been tested, and some are known to be poisonous.

When and where to find it. The fruit bodies are usually solitary at the line between the mineral soil and the duff, and are found mostly while digging for false truffles. It fruits during the summer and early fall in the white-bark pine region of Idaho. During the rainy season the fruit bodies sometime push through the duff to at least partly expose the cap.

Microscopic characters. **Spores** 9–11 x 6–7 μ, verrucose-rugulose, inequilateral in profile, ovate in face view. **Pileus trama** covered by a thin gelatinous pellicle of brownish (in KOH) narrow appressed hyphae. **Clamps** present.

Field identification marks. (1) The bright bluish lavender color of the cap; (2) the heavy aromatic (subfetid) odor; (3) the sheath on the stalk matching the color of the cap; (4) the cap is sticky when fresh.

Observations. This is one *Cortinarius* that can be recognized at a glance in the field.

Edibility. Not known.

When and where to find it. Scattered under conifers — hemlock, Douglas fir, and species of *Abies* (true firs), in the fall, never in great abundance. It is most abundant on the Upper Priest River in Idaho, which may mean that it is more abundant in southwestern Canada than in the United States.

Microscopic characters. **Spores** 7–9 (10) x 5–5.5 μ, rusty brown, warty-rugulose, in face view subelliptic, in profile slightly inequilateral. **Basidia** 4-spored. **Pleuro-** and **cheilocystidia** none seen. **Pellicle of pileus** of narrow subgelatinous hyphae, the layer a delicate pink as revived in KOH. **Clamps** present.

132 Cortinarius percomis

Field identification marks. (1) The penetrating sweetly aromatic odor; (2) a pale bright yellow to bright ochraceous viscid cap; (3) sulphur yellow outer veil.

Observations.   The description fails to convey the beauty of this species. There are many yellow species of sticky *Cortinarii* but only one with the odor of this one.

Edibility.   No data on it.

When and where to find it.   Gregarious to cespitose in the spruce-fir zone of our western mountains, especially in the Olympic National Park during the fall.

Microscopic characters.   **Spores** 10–13 x 5.5–6.5 μ, warty-rugulose, in profile inequilateral, ovate in face view, rusty brown in KOH. **Basidia** 4-spored, when revived in KOH some containing purplish to vinaceous granules. **Pileus** with a pellicle of gelatinous hyphae. **Clamps** present.

133 Cortinarius crassus

Field identification marks.   (1) The cap is sticky, clay color darkening to russet, and very fleshy; (2) the stalk is short and thick; (3) the gills are white at first, dark cinnamon in age; (4) the stalk is pallid slowly becoming dingy brown from handling.

Observations.   A second species with about the same stature is *C. subbalteatus* with a violaceous cap margin. A third is *C. largus* which at first has violaceous gills, but in age may be mistaken for *C. crassus*. All three occur in our western area.

Edibility.   Not recommended. If the fungus described here is found to be edible and tasty it would be an important species for the western pot hunter. However, various accounts by different authors indicate that more than one species is passing under this name.

When and where to find it.   Solitary to gregarious in mountain conifer forests, especially in the Olympic National Park, during the fall.

Microscopic characters.   **Spores** 9–12 x 5.5–6.5 μ, roughened, inequilateral in profile, ovate in face view, tawny in KOH. **Basidia** 4-spored. **Cystidia** none. **Pileus** with a thin gelatinous pellicle of yellow brown hyphae (in KOH). **Clamps** present.

132 *Cortinarius percomis*   *About two-thirds natural size*

133 *Cortinarius crassus*   *About one-half natural size*

134　Cortinarius turmalis

Field identification marks.　(1) The fulvous sticky cap; (2) the white stalk with an apical annular zone where the veil breaks; (3) lack of any distinctive odor or taste; (4) whitish gills when young.

Observations.　In our western area this species is often very robust, but regardless of the size of the fruit bodies, it is distinct from *Cortinarius triumphans* and related variants because of the lack of yellow veil remnants on the stalk below the annular zone.

Edibility.　Not recommended.

When and where to find it.　Scattered to gregarious-cespitose during late summer and in the fall in conifer forests. It is one of the common *Cortinarii* in our western mountains. The best fruitings seen have been in the vicinity of Olympic Hot Springs in the Olympic National Park.

Microscopic characters.　**Spores** 8–10 x 5–6 μ, minutely roughened, ovate to subelliptic in face view, in profile slightly inequilateral. **Basidia** 4-spored. **Cystidia** none. **Cuticle** of pileus a pellicle of appressed gelatinous hyphae yellow in KOH and with clamps at the cross walls.

135　Cortinarius rubicundulus

Field identification marks.　(1) The overall stature; (2) the whitish to dull pale tan to yellow to reddish progression of color from youth to age over the fruit body; (3) the cortina and the fibrillose unpolished cap. The cystidia are a valuable microscopic character.

Observations.　This species was first reported for North America under the name *C. pseudobolaris*. The latter is now regarded as a synonym of *C. rubicundulus*.

Edibility.　Not enough is known about it to rate it in North America.

When and where to find it.　Like most *Cortinarii* it does not fruit in an even pattern from year to year, and during "peak" years, though frequently encountered, is not found in quantity. From my experience it appears to be chiefly a coastal species in our area, and its best locality to date is the Cascade Head Experimental Forest at Otis on the Oregon coast.

Microscopic characters.　**Spores** 6–7.5 x 4–4.5 μ, broadly inequilateral in profile, elliptic in face view, medium tawny in KOH, verruculose. **Basidia** 4-spored, hyaline to dingy reddish cinnamon in KOH (fading on standing). **Pleuro-** and **cheilo-cystidia** similar and numerous, 34–48 x 7–14 μ, cylindric to clavate or ventricose, apex often proliferated, hyaline to dingy cinnamon in KOH. **Cuticular layer** of pileus indefinite (hyphae about like those of the context). **Clamps** present.

34 *Cortinarius turmalis*   *About one-half natural size*

35 *Cortinarius rubicundulus*   *About two-thirds natural size*

Field identification marks.   (1) The lilac cap and stalk; (2) the croceus to rusty cinnamon interior of cap and stalk; (3) the pungent faint odor; (4) a dry fibrillose cap.

Observations.   This is a most interesting species in North America, where it seems to be more variable than in Europe. In the Priest Lake district of Idaho we have a population widely distributed in which the cap shows sectoring — whitish areas in a wedge-shaped pattern (broadest at the margin of the cap). The typical form occurs commonly in the Cascade and Olympic mountains. This species is one of the best examples to show the localization of pigment to different parts of the fruit body, a feature very important in the recognition of species of *Cortinarius*.

Edibility.   I have no data on it, and do not recommend it.

When and where to find it.   Solitary to gregarious in old-growth hemlock and spruce-fir forests during the fall months in the Pacific Northwest, not uncommon but the sectoring variant is known to date only from northern Idaho.

Microscopic characters.   **Spores** 7–9 x 4.5–5.5 μ, warty-rugulose, ellipsoid to subovoid. **Basidia** 30–40 x 8–12 μ, clavate, hyaline in KOH. **Pleurocystidia** present as pseudocystidia with brown amorphous content, 27–44 x 6–10 μ; no leptocystidia observed. **Cheilocystidia** present as hyphal end cells 20–40 x 4–6 μ, hyaline, thin-walled, typically cylindric with blunt apex. **Gill trama** parallel, hyphae narrower toward subhymenium. **Pileus cuticle** of hyphae 5–12 μ wide in ± radial arrangement, walls smooth, thin and hyaline. **Clamps** present. Pigment deposits none.

136 *Cortinarius traganus*   *About one-half natural size*

About one-half natural size

Cortinarius armillatus 137

Field identification marks. (1) The dull red band or bands on the stalk from the outer veil; (2) the subhygrophanous cap (one may have difficulty deciding in which subgenus the species belongs); (3) pale dingy cinnamon gills when young; (4) an odor somewhat resembling that of radish.

Observations. For accurate identification one should add the large spores to the above set of features. This species is found in northern Idaho where birch is not uncommon in the cut-over areas. West of the Cascades a somewhat similar species in the conifer forests is *C. haematochelis*. It has smaller spores.

Edibility. Edible.

When and where to find it. Solitary to gregarious or in clusters, in areas where birch *(Betula)* occurs. It is most abundant (but not common) in the Priest Lake district of Idaho in the fall.

Microscopic characters. **Spores** 10–12 (13) x 5.5–7 (7.5) μ, warty-rugulose, slightly inequilateral in profile, elliptic in face view, rusty brown in KOH. **Basidia** 4-spored. **Pleurocystidia** none. **Cheilocystidia** as flexuous filaments 3–4 μ wide, projecting slightly. **Pileus** with a thin pellicle of nongelatinous narrow hyphae. **Clamps** present.

138 *Cortinarius californicus*   *Slightly less than natural si*

139 *Cortinarius subaustralis*   *About one-half natural siz*

Field identification marks. (1) The hygrophanous ferruginous red cap; (2) the orange red gills; (3) the dull orange to paler stalk decorated with remnants of the orange cortina; (4) the orange red mycelium at base of stalk.

Observations. The hygrophanous (one color when moist, paler when dry) cap separates this species from numerous brightly colored species of the subgenus *Dermocybe* in which the cap is dry and fibrillose.

Edibility. Not tested.

When and where to find it. Scattered under spruce and fir during the fall season in Oregon and northern California, uncommon.

Microscopic characters. **Spores** 7–9 x 4–4.5 μ, ellipsoid to ovoid, warty-rugulose, dark rusty brown in KOH under the microscope. **Basidia** 4-spored. **Pleurocystidia** and **cheilocystidia** not differentiated. **Clamps** present.

Field identification marks. This species cannot be accurately identified in the field. Reference to the pseudocystidia must be included. The diagnostic characters are: (1) the fibrillose brown cap; (2) white gills at first; (3) lack of color changes when bruised; (4) spores 6–7.5 x 3.5–4 μ; (5) the presence of pseudocystidia.

Edibility. Not known.

When and where to find it. This species was described from North Carolina but has turned up in the conifer forests of our western area. I suspect an association with hemlock. It appears in late summer and fall, solitary to gregarious in forests where hemlock is present. It may well be more common than we think at present.

Microscopic characters. **Spores** 6–7.5 x 3.5–4 μ, very pale brown under the microscope, nearly smooth, somewhat inequilateral. **Basidia** 4-spored. **Pleurocystidia** (or pseudocystidia) abundant to scattered, 50–70 x 8–10 μ, subcylindric to subfusiform, originating from laticiferous hyphae in the trama. **Cheilocystidia** similar to pleurocystidia only many of them with proliferated apex. **Cuticle** of pileus scarcely different from the trama, the hyphae near the surface with tawny walls. **Clamps** present.

Field identification marks.   (1) Habitat is on naked sand with lodgepole pine usually within a radius of 50 feet from the fruit bodies; (2) the gills are soon orange to brownish orange and the cap a dark reddish brown; (3) the cap at maturity is appressed fibrillose with the fibrils arranged in streaks; (4) the fruit bodies barely project above the level of the sand; (5) they are often numerous in a small area.

Observations.   The spores at once distinguish this species from others in the subgenus *Dermocybe.* It forms mycorrhiza with lodgepole pine, in particular the race occurring in the sand dunes along the coast.

Edibility.   To my knowledge, it has not been tested. It is next to impossible to remove all the sand, and the fruit bodies are small so it is not likely to attract the "pot hunter."

When and where to find it.   See field identification marks, above. It is found late in the fall season often in abundance after heavy rains.

Microscopic characters.   **Spores** 9.2–13 x 3.5–4.5 (5) μ, smooth or surface merely slightly irregular, in profile obscurely to somewhat inequilateral, in face view narrowly elliptic to ovate. **Pleurocystidia** absent. **Cheilocystidia** 14–31 x 6–8 μ, clavate or nearly so. Interhyphal pigment deposits present in revived material and ochraceous or orange in KOH. **Clamps** present.

Rozites

141   Rozites caperata

Field identification marks.   (1) The rusty brown spore deposit; (2) the ring on the stalk formed by the partial veil; (3) the stout stalk; (4) a pallid silky coating often most evident around the center of the cap.

Observations.   This species is often characterized as "a *Cortinarius* with a membranous annulus." However, there are *Cortinarii* with a ring formed by an outer veil, so one must be careful. Fortunately *R. caperata* has a distinctive appearance which once seen is usually remembered. One should collect this species a number of times before eating it just to be sure of its identity.

Edibility.   Edible and choice. Discard the stalks as they are tough.

140 *Cortinarius aureifolius*

Twice natural size

141 *Rozites caperata*

About one-half natural size

When and where to find it. It is a common species in conifer country, but occurs under hardwoods as well. It fruits in the fall and is abundant during warm wet seasons.

Microscopic characters. **Spores** 12–14 x 7–9 μ, in profile somewhat inequilateral, verrucose. **Pleurocystidia** none. **Pileus cuticle** a thick pellicle of subgelatinous hyphae. **Clamps** present.

Phaeocollybia

Key to Species

1.   Gills dull lilac when young . . . . (p. 190) *P. fallax*
1.   Gills whitish becoming brown . . . . (p. 190) *P. olivacea*

---

142   Phaeocollybia fallax

Field identification marks.   (1) The slimy, olive, glabrous cap;
(2) lilaceous to caesious gills; (3) the slender stalk; (4) possible association with Sitka spruce.

Observations.   As yet I have not been able to determine the
point or object from which the pseudorhiza originates. The
pseudorhiza tapers to a fine thread which is easily broken
when one tries to trace it through the mass of spruce rootlets.

Edibility.   Not known.

When and where to find it.   Solitary to gregarious in the fall
in habitats containing Sitka spruce along with other species
of trees. Not uncommon during warm wet rainy seasons along
the coast of Washington, Oregon, and northern California.

Microscopic characters.   **Spores** 7–9 x 4.5–5.5 μ, verruculose,
inequilateral in profile, ovate and beaked as seen in face view,
rusty brown in KOH. **Basidia** 4-spored. **Cheilocystidia** clavate,
mostly resembling large basidioles. **Pileus** with a cuticle in
the form of a tangled turf of gelatinous hyphae 2–3 μ wide.
Clamps present.

---

143   Phaeocollybia olivacea

Field identification marks.   (1) The slimy olive to olive brown
cap; (2) the olive context; (3) the pallid (whitish) gills when
young; (4) the stalk 10–20 mm thick near apex; (5) watery
olivaceous above and rusty orange near or below the ground
line; (6) the habit of fruiting in large groups of 15–50 fruit
bodies.

Observations.   It has not been demonstrated what the
substratum is for this species because it is so difficult to trace
the pseudorhiza to its point of origin.

Edibility.   Not known at least to the writer, and I would not
encourage experimenting with it.

When and where to find it.   It appears to be a coastal species fruiting late in the fall and extending south from northern Oregon. It is very abundant in southern Oregon. During
poor seasons only solitary to scattered fruit bodies can be
found.

Microscopic characters.   **Spores** 8–11 x 5–6 μ. **Cheilocystidia**
clavate to filamentous, 24–35 x 4–10 μ. **Pileus cuticle** an
ixotrichoderm, hyphae 2–4 μ wide. **Clamps** present.

2 *Phaeocollybia fallax*   *Slightly less than natural size*

3 *Phaeocollybia olivacea*   *About two-thirds natural size*

191

Hebeloma

Key to Species

1.	Stalk scaly, but no veil present on young fruit bodies; cap dull vinaceous brown . . . . (p. 192) *H. sinapizans*
1.	Stalk smooth, no veil present on young fruit bodies; cap whitish to buff-colored . . . . (p. 192) *H. crustuliniforme*

## 144	Hebeloma crustuliniforme

Field identification marks. (1) The crushed flesh has a pungent odor resembling that of radish; (2) the cap is sticky and pallid to a pale dingy alutaceous; (3) no veil of any kind is present; (4) the stature is medium sized; (5) the gill edges are often beaded with drops of a liquid.

Observations. Species of *Hebeloma* are difficult to distinguish, but the features listed above should enable an accurate identification to be made taken in conjunction with the photograph.

Edibility. Reported to be poisonous.

When and where to find it. Scattered to gregarious under conifers, and also under hardwoods, late summer and fall depending on when the rains come. It is a rather common species in the western area.

Microscopic characters. **Spores** 9–12 x 5–7 μ, smooth, ovate to elliptic in face view, clay brown in KOH. **Basidia** 4-spored. **Pleurocystidia** none. **Cheilocystidia** 24–40 x 6–9 μ, cylindric to narrowly clavate, hyaline, abundant. **Pileus** with a thin gelatinous pellicle. **Clamps** present.

## 145	Hebeloma sinapizans

Field identification marks. (1) The stalk covered with distinct scales, *but* no veil is present — check young fruit bodies; (2) a strong pungent odor and taste reminding one of radishes; (3) cap surface slightly sticky when fresh and a dull medium brown color, but usually with an overlying pallid sheen; (4) stalk 1–3 cm thick (a robust fruit body).

Observations. This species may be mistaken for a *Cortinarius* by many, but no veil is present even on the button stages.

Edibility. Not recommended. Poisonous species are known in this genus, and distinguishing the species from each other is very difficult. As yet we have no truly definitive work on the North American species.

When and where to find it. Gregarious to gregarious-cespitose, often in large numbers during warm wet fall weather, usually in mixed conifer-hardwood forests, especially mixtures of birch and balsam *(Betula* sp. and *Abies* sp.).

44 *Hebeloma crustuliniforme*   *About two-thirds natural size*

45 *Hebeloma sinapizans*   *About one-half natural size*

Microscopic characters. **Spores** 10–12.5 x 6–7 μ, in profile ± inequilateral, in face view subovoid, rugulose-warty, pale cinnamon in KOH. **Basidia** 4-spored. **Pleurocystidia** absent. **Cheilocystidia** abundant, 48–70 x 10–12 μ, clavate to pedicellate-subcapitate. **Gill trama** of parallel to subparallel hyphae. **Cuticle** of pileus a thick gelatinous pellicle. **Clamps** present.

146 Gymnopilus terrestris

Field identification marks. (1) The lack of a veil; (2) the essentially terrestrial habitat; (3) the mild odor and taste; (4) the ferruginous to orange ferruginous cap; (5) the dull orange ochraceous gills when young.

Observations. This is the fungus previously referred to as *Naucoria sticticus,* a species, it now appears, still very poorly understood in Europe. Most collectors will mistake it for a *Cortinarius* in which the cortina has been obliterated.

Edibility. Not known. Experimentation definitely discouraged.

When and where to find it. Solitary to gregarious in conifer forests throughout the area during the fall season but seldom in quantity.

Microscopic characters. Spores 5.5–7 x 4–5.5 (6) μ, ovoid to elliptic (rarely subglobose) in face view, in profile slightly inequilateral, verruculose, ferruginous in KOH, dextrinoid. **Cuticle** of pileus a layer of appressed hyphae, some with incrustations and the end cells often ascending as pileocystidia, the latter 33–50 x 6–10 μ. **Clamps** present.

Inocybe

Key to Species

1. Fruit body white staining reddish .... (p. 196) *I. pudica*
1. Fruit body ochraceous brown .... (p. 194) *I. olympiana*
    (NOTE: *About half the species in this genus will key out here, but all are to be regarded as poisonous whether they have been tested or not.)*

147 Inocybe olympiana

Field identification marks. Species of *Inocybe* are not identified at sight in the field. Both field and microscopic characters are essential. For the present species the characteristics are: (1) The ochraceous brown appressed-fibrillose cap; (2) the stalk tawny overall in age; (3) spores 7–9 x 4–5 μ; (4) the very abundant, long pleurocystidia (60–90 x 10–16 μ) with their walls thickened and yellow in KOH.

Observations. A closely related species is *I. subochracea.*

Edibility.   Dangerous. All *Inocybe* species fall in this category or are in the definitely poisonous group. DO NOT experiment with any in this genus.

When and where to find it.   Gregarious in the conifer forests of the region, especially in the Olympic National Park, during the fall collecting season.

Microscopic characters.   **Spores** 7–9 x 4–5 μ, smooth, inequilateral in profile. **Basidia** 4-spored. **Pleurocystidia** 60–90 x 10–16 μ, fusoid-ventricose to subcylindric, thick-walled, walls yellow in KOH, apex incrusted. **Cheilocystidia** similar to pleurocystidia but shorter. **Clamps** present.

46 *Gymnopilus terrestris*   *Slightly less than natural size*

47 *Inocybe olympiana*   *About natural size*

148   Inocybe pudica

Field identification marks.   (1) Fruit body white overall at first but soon stained or flushed pink to red; (2) odor of crushed flesh, and rather disagreeable taste; (3) the cap is not scaly but rather appressed fibrillose and wet to tacky to the touch but not truly viscid (sticky); (4) gills white at first but clay color to dull cinnamon at maturity; (5) a fibrillose veil is present on young fruit bodies.

Observations.   This species might be confused with *Cortinarii* such as *C. uliginosus,* but microscopic characters such as thick walled cystidia and smooth spores distinguish it from any *Cortinarius.*

Edibility.   A safe rule to follow is to avoid *all* of the ± four hundred species of *Inocybe* known for North America. It has been shown by chemical analysis that *many* of them contain relatively large amounts of poison.

When and where to find it.   Most collectors will find this species on their first collecting trip if they visit pole-size stands of Douglas fir. It is one of the commonest species in the West. It fruits from late summer to winter depending on weather and location.

Microscopic characters.   **Spores** clay color in deposit, 7–9 x 4.5–5.5 μ, smooth, ± elliptic in face view, in profile slightly inequilateral, pale tawny in KOH. **Basidia** 4-spored. **Pleurocystidia** 45–60 x 10–18 μ, walls somewhat thickened. **Cheilocystidia** ± as for pleurocystidia. **Cuticle** of pileus not sharply differentiated as a layer. **Clamps** present.

Galerina

Galerina venenata   149

Field identification marks.   (1) The cap is pale bay brown to reddish cinnamon and glabrous; (2) the gills are a golden tawny (rather dull in age); (3) the stalk bears a thin bandlike ring above the middle in most fruit bodies; (4) the taste of the raw context is slowly disagreeable and leaves a burning sensation in the throat; (5) the fruit bodies appear to be terrestrial.

Observations.   Microscopic features are necessary to make an accurate identification because a number of other species are quite similar in appearance.

Edibility.   POISONOUS: Under no circumstances should any one experiment with this species or any resembling it. It is very poisonous, the poisons being of the *Amanita* type.

When and where to find it.   It appears to be a fall fruiting fungus and is gregarious and terrestrial, but, where I have

149 *Galerina venenata*   *About natural size*

seen it, most of the other fungi in the lawn at the time were species of wood-inhabiting fungi. Also, the *Galerina* species most closely resembling it grow on wood. It occurs west of the Cascades in Washington and Oregon.

Microscopic characters. **Spores** 8.7–11.5 x 5–7 μ (basidia 4-spored), 10–16 x 5.5–8 μ (basidia 2-spored), surface strongly rugose-verrucose, deep rusty brown in KOH. **Pleurocystidia** 35–67 x 10–15 μ, ventricose but generally variable in shape, apex subacute to capitate. **Cheilocystidia** resembling pleurocystidia more or less. **Pileus cuticle** not sharply differentiated. **Clamps** present.

150   Galerina paludosa

Field identification marks.   (1) The ochraceous tawny cap; (2) the habit of fruiting scattered to gregarious on sphagnum; (3) the copious remains of a veil on the stalk often leaving a superior ring; (4) the stalk 5–10 (20) cm long.

Observations.   This is a characteristic species of sphagnum bogs throughout the Northern Hemisphere. The depth of the moss has a direct bearing on the length of the stalk. It has the most copiously developed veil of any *Galerina* on *Sphagnum* (but some species of *Cortinarius* might possibly be confused with it).

Edibility.   Not recommended. The genus is dangerous.

When and where to find it.   Scattered to gregarious in sphagnum bogs over the Northern Hemisphere during the late spring through to fall, common in its habitat. I have seen it mostly in Idaho, Washington, and Oregon in the West.

150 *Galerina paludosa*   *Slightly less than natural size*

Microscopic characters. **Spores** (8) 9–11 x 5–7 (8) μ, in face view broadly ovate, in profile slightly inequilateral, punctate-rugulose in KOH, plage area distinct. **Basidia** 4-spored. **Pleurocystidia** none. **Cheilocystidia** (25) 30–45 x 6–12 x 3–5 x 3.5–9 μ, ventricose near the pedicel, neck narrow, apex somewhat enlarged. **Pileus** lacking a distinct cuticle, the hyphae near the surface 8–15 μ wide. **Clamps** present.

Pholiota

The genus is here treated in the sense of Smith and Hesler (1968), which is to say that it includes *Kuehner-omyces,* most of the smooth-spored species of *Flammula* in the Friesian sense, some brown-spored species previously referred to *Naematoloma,* and a number from *Phaeomarasmius.* The species are among the more conspicuous wood-inhabiting mushrooms.

Key to Species

151 *Pholiota terrestris*   *Slightly less than natural size*

Field identification marks. (1) Densely clustered and appearing to be terrestrial; (2) the dark dull brown fibrillose-scaly cap; (3) the gills pallid becoming avellaneous (grayish brown) and in age with a yellow tinge; (4) the stalk scaly with dull brown scales similar to those on the cap; (5) cap not obviously sticky to the touch.

Observations. The size of the cluster and the size of the fruit body varies greatly from season to season.

Edibility. Edible, so I am informed by a number of collectors. In view of recent events relative to poisonings by *Pholiota,* experimentation with this genus is not encouraged.

When and where to find it. It fruits in late summer and fall in lawns (near trees) and along roads on packed soil. It appears to be terrestrial, but is almost certainly living on buried wood, in fact a few collections were attached to wood. It is a common species in the Northwest in the fall season.

Microscopic characters. **Spores** 4.5–6.5 (7) x 3.5–4.5 μ, smooth, apical pore distinct but small, elliptic in face view, in profile subelliptic to slightly bean-shaped. **Basidia** 4-spored. **Pleurocystidia** 18–34 x (4) 5–10 (12) μ, clavate, mucronate or fusoid-ventricose, with a refractive inclusion as revived in KOH. **Cheilocystidia** 26–50 x 4–8 μ, numerous, cylindric, subcapitate, subutriform to fusoid-ventricose, thin-walled. **Pileus cuticle** with an epicutis of nongelatinous brown hyphae with incrusted walls, the cells inflated to 12 μ at times; subcutis a narrow gelatinous layer of brown hyphae with incrustations on the walls. **Clamps** present.

Field identification marks. (1) The coarse dull rusty brown spikelike scales which cover the young cap; (2) the very scaly stalk; (3) the whitish gills when young; (4) a thin sticky layer beneath the scales of the cap which shows best at maturity or in rain-washed specimens.

Observations. This is one of the conspicuous species in the genus in the Northwest. The pale form of *Pholiota squarrosa* may be mistaken for it, but lacks the gelatinous layer in the cap and is more abundant in the southern Rocky Mountains than the Northwest.

Edibility. Edible and choice, but *P. squarrosa* which is *usually* thought to be harmless has caused cases of poisoning. They have been of the "mild" type. In other words, any one collecting for the table should be sure he knows his species and should observe the usual precautions.

When and where to find it. Clustered on hardwood logs and stumps during late summer and fall, common and abundant during many seasons, but sometimes failing to fruit.

Microscopic characters. **Spores** 4–5.5 (6) x (2.5) 3–3.5 μ, smooth, ovate to elliptic in face view, in profile subelliptic to obscurely inequilateral, dull cinnamon in KOH. **Pleurocystidia** abundant (25) 30–50 (65) x 8–15 (18) μ, clavate, mucronate or fusoid-ventricose, often with an apical elongation, content opaque in KOH but lacking amorphous refractive inclusions. **Cheilocystidia** 26–40 (60) x 5–11 (13) μ, clavate to fusoid-ventricose. **Pileus cuticle** consisting of a gelatinous subcutis and an epicutis of erect scales composed of hyphae having isodiametric cells, with smooth to incrusted walls, brown pigment located in the wall. **Clamps** present.

152 *Pholiota squarrosoides*   About two-thirds natural size

152 *Pholiota squarrosoides*   About two-thirds natural size

Field identification marks. (1) The zones of dry tawny scales low down on the stalk which represent outer veil remnants; (2) the slimy cap with large tawny spotlike gelatinous scales (in wet weather); (3) the habit of growing primarily on wood of deciduous trees.

Observations. The relatively large spores with a well-developed apical pore are an important microscopic feature. *P. aurivelloides* has even larger spores furnished with a truncate apex from the broad apical pore. The latter is more frequent, apparently, in the southern Rocky Mountains.

Edibility. Edible. I do not recommend it for collectors in western North America (see the poisonous *P. hiemalis).*

When and where to find it. It fruits during the summer and fall in clusters, or solitary, typically on remains of deciduous trees. As yet, because of previous confusion of taxa in this group, an accurate picture of its distribution in our western area cannot be given. I have not found it fruiting in abundance.

Microscopic characters. **Spores** 7–9.5 (11) x 4.5–6 μ, smooth, apical pore distinct but apex not truncate, broadly elliptic in face view, in profile subelliptic to bean-shaped. **Basidia** 4-spored. **Pleurocystidia** 30–45 x 4–8 μ, mostly fusoid (some branched near the apex, giving the impression of 3–4 sterigmata), content hyaline to partly filled with refractive amorphous material. **Cheilocystidia** 26–35 x 5–10 μ, subfusoid **Pileus** with a thick gelatinous pellicle of interwoven yellowish hyphae 2–5 μ wide. **Clamps** present.

153 *Pholiota aurivella*

About natural size

154 Pholiota hiemalis (The Winter Pholiota)

Field identification marks. (1) The slimy yellow cap covered by a number of broad gelatinous scales; (2) the scattered rusty brown gelatinous scales on midportion of the stalk; (3) the fibrillose annular zone near the apex of the stalk; (4) habitat on conifer logs

Observations. This species has the slimy veil of *P. adiposa* but its spores are much larger. As pointed out by Smith and Hesler, *P. adiposa* is actually exceedingly rare in North America. *P. abietis* and *P. filamentosa* also occur on wood of conifers in the Pacific Northwest. Compare the descriptions given in the present work

Edibility. POISONOUS. The only report we have on this is one authenticated by Dr. E. Tylutki of the University of Idaho who recognized the species as the one which caused a case of mushroom poisoning which came to his attention. Because of this, I no longer recommend the *P. aurivella* group of species for the "pot hunter."

When and where to find it. Cespitose-gregarious on rotting conifer logs during cool or cold wet weather late in the fall. It occurs in great abundance during some seasons in northern Idaho, and withstands considerable freezing.

Microscopic characters. **Spores** 7–9 (10) x 4–4.5 (5) μ, smooth, ellipsoid or nearly so, with a minute pore at the apex (under oil-immersion). **Pleurocystidia** 30–50 x 8–15 μ, clavate-mucronate to clavate. **Cheilocystidia** clavate to vesiculose, 26–40 x 9–13 μ. **Subhymenium** gelatinous. **Pileus cuticle** a thick layer of gelatinized hyphae 3–7 μ wide and in a tangled arrangement. **Clamps** present.

154 *Pholiota hiemalis*   *About natural size*

Field identification marks. In the stirps *Adiposa* in the work of Smith and Hesler (1968) this species stands out by: (1) its habitat on conifer wood; (2) the presence of a distinct membranous ring on some of the fruit bodies in a collection; (3) a greenish yellow to lemon yellow cap in contrast to the duller yellow of many others in the stirps.

Observations. The small spores clearly distinguish it from *P. aurivella* should there be any question concerning the other features. *P. squarrosa-adiposa* has the same spores but lacks a membranous ring and grows on hardwoods mainly. (Compare *P. abietis* and *P. hiemalis* also.)

Edibility. Not recommended.

When and where to find it. Its habitat is much the same as that for *P. abietis*. It has been found most frequently in the Priest Lake district of Idaho in the fall.

Microscopic characters. **Spore deposit** rusty brown. **Spores** 6–7.5 (8) x 3.8–4.2 μ, smooth, apical pore minute; in face view elliptic, in profile elliptic to obscurely bean-shaped. **Basidia** 4-spored. **Pleurocystidia** 25–40 x 6–13 μ, narrowly clavate-mucronate to broadly so, pedicel long, content as for that of chrysocystidia, (but rusty brown in some in KOH). **Cheilocystidia** 18–32 x 5–11 μ, clavate to fusoid-ventricose. **Pileus** with a gelatinous pellicle of hyphae 3–6 μ wide, hyaline to yellow in KOH and more or less appressed. **Clamps** present.

155 *Pholiota filamentosa*   *Slightly less than natural size*

Field identification marks.　(1)The brownish veil remnants mostly near the margin of the cap; (2) the subcespitose to cespitose habit, and usually appearing to be terrestrial; (3) the generally orange fulvous to ochraceous orange tones of the cap; (4) the pallid young gills.

Observations.　For an accurate identification one needs to consult the technical literature. The *"Flammula*-like" species of *Pholiota* in our western flora are numerous and confusing.

Edibility.　Not known.

When and where to find it.　Subcespitose to cespitose and appearing terrestrial (though often on or close to very decayed conifer wood) late summer in the Rocky Mountains and during the fall season west of the Cascades. Frequently associated with remains of lodgepole pine stumps and fallen trunks.

Microscopic characters.　**Spores** 5.5–7 x 3.2–4 μ, smooth, oblong to ovate in face view, in profile obscurely to distinctly bean-shaped. **Basidia** 4-spored. **Pleurocystidia** abundant, 45–83 x 9–16 μ, fusoid-ventricose, apex obtuse to subacute, content (in KOH) evenly ochraceous, wall up to 0.5 μ thick as revived in KOH. **Cheilocystidia** 32–50 x 8–12 μ, shape as for pleurocystidia. **Cuticle** of pileus a thick gelatinous layer of hyphae 2–3 μ thick. **Clamps** present.

156 *Pholiota sublubrica*　　　　　　　　　　　　*Slightly less than natural size*

Field identification marks. (1) Growing on wood of conifers; (2) cap with spotlike tawny brown gelatinous appressed scales; (3) gills pallid brownish when young; (4) a distinct membranous ring not present on the stalk (merely a fibrillose annular zone where the veil breaks); (5) the stalk not darkening appreciably over the basal area.

Observations. This species differs from *P. hiemalis* in lacking gelatinous veil remnants on the stalk, and in having smaller spores.

Edibility. Definitely not recommended. See discussion of *P. hiemalis.*

When and where to find it. Typically clustered on dead fir logs or stumps in the fall. It appears that this species is widespread in the West but was long confused with *"Pholiota adiposa,"* a species misidentified on this continent for the last 75 years by professionals and amateurs alike.

Microscopic characters. **Spores** 5.5–7 (7.5) x 3.5–4 (4.5) μ, smooth, oblong to elliptic in face view, subelliptic in profile, apical pore minute. **Basidia** 4-spored. **Pleurocystidia** 32–46 x 8–12 μ, clavate to clavate-mucronate, with dark reddish brown content as revived in KOH, or some with a refractive amorphous inclusion paler in color (as in chrysocystidia of many authors). **Cheilocystidia** (23) 32–56 x (4) 6–9 (11) μ, versiform but mostly fusoid-ventricose to subcylindric-capitate, hyaline in KOH (yellow to tawny if specimen was poorly dried). **Pileus** with a thick gelatinous cuticle reddish in KOH (as revived), the hyphae 4–10 μ wide and appressed. **Clamps** present.

157 *Pholiota abietis*   *Slightly less than natural size*

158 Pholiota subcaerulea

Field identification marks. (1) The blue to greenish blue color of the cap and stalk; (2) the dull rusty brown gills when mature; (3) the ring not membranous and persistent; (4) the slimy-sticky cap.

Observations. This and *Stropharia aeruginosa* differ mainly in the color of the spore deposit. Both occur in the West but according to my experience *S. aeruginosa* must be rare in this region.

Edibility. Not recommended.

When and where to find it. In small clusters in long grass under Douglas fir, alder, and dogwood in the fall. Its distribution and abundance in the region remains to be studied. At present it is known from Idaho, Washington, and Oregon.

Microscopic characters. **Spores** 7–9 x 4–4.5 μ, smooth, ovate to subelliptic in face view; obscurely inequilateral in profile, apex not truncate (pore minute). **Basidia** 4-spored. **Pleurocystidia** 24–36 x 9–12 μ, clavate, clavate-mucronate, or fusoid-ventricose, hyaline but with a refractive body as revived in KOH, in Melzer's the refractive body bright red to bay red (very conspicous). **Cheilocystidia** variable: (1) 30–52 x 2.5–4 x 5–8 μ, narrowly subcapitate-pedicellate to narrowly clavate, hyaline in KOH; (2) 20–28 x 4–7 μ, fusoid, hyaline; (3) resembling pleurocystidia. **Caulocystidia** clavate to sub-filamentose, 4–6 μ wide at apex. **Pileus cuticle** a thick gelatinous pellicle of interwoven hyphae 2–5 μ wide, hyaline, smooth. **Clamps** present.

159 Pholiota albivelata

Field identification marks. (1) The glabrous viscid vinaceous brown cap; (2) the broad, membranous ring which is white and persistent; (3) the base of the stalk with numerous white rhizomorphs (strands of mycelium); (4) gills grayish brown when mature, white at first; (5) usually growing solitary or widely scattered.

Observations. There are two species in this group: *P. albivelata* with spores 7–9 x 4–5.5 μ, and *P. sipei* with spores 9–12 x 4.5–6 μ. In *P. sipei* the gills are cinnamon colored when mature, and it apparently is rather rarely collected. *P. albivelata* is rather common and widely distributed throughout the Pacific Coast area.

Edibility. Not recommended. I have no data on it, and since some of the *Pholiota* species have turned out to be poisonous, caution is the word.

When and where to find it. Solitary to scattered on debris under conifers in the fall throughout the coastal area of Washington, Oregon, and northern California. I find it on numerous occasions but very rarely more than a few fruit bodies at a time.

Microscopic characters. **Spore deposit** rusty brown ("cinnamon brown"). **Spores** 7–9 x 4–5.5 μ, smooth, apical pore very minute; shape in face view elliptic to ovate, in profile subelliptic, dull cinnamon under the microscope in KOH. **Basidia** 4-spored. **Pleurocystidia** abundant, 30–60 x 5–12 μ, clavate-mucronate, with a refractive coagulated content as revived in KOH (they are chrysocystidia). **Cheilocystidia** 20–56 x 3–7 μ, filamentose-subcapitate. **Caulocystidia** resembling cheilocystidia but scattered to rare. **Subhymenium** a gelatinized layer. **Pileus epicutis** a thick (100 μ) gelatinized layer of hyphae 1.5–3 μ wide. **Clamps** present.

158 *Pholiota subcaerulea*   About natural size

159 *Pholiota albivelata*   Slightly less than natural size

160 Pholiota mutabilis (The Changeable Pholiota)

Field identification marks. (1) The stalk is covered by fine recurved scales up to the ring or annular zone; (2) the cap is glabrous, and very hygrophanous; (3) it grows in dense clusters on wood of broad-leaved trees as well as on wood of conifers, and is fairly persistent; (4) the cap has a thin separable covering.

Observations. This species should not be mistaken for a *Galerina* by anyone paying attention to the above mentioned features and the photograph. *Galerina autumnalis* has a thin bandlike ring on the stalk, lacks recurved squamules below it, and does not ordinarily occur in large clusters. If a microscope is available the two are readily distinguished since the spores of the *Galerina* are roughened.

Edibility. *P. mutabilis* is recognized as a good edible species, but one must be *very careful* to avoid confusing it with *Galerina autumnalis*. Hence *P. mutabilis* is not recommended to those lacking considerable experience in the identification of higher fungi.

When and where to find it. It occurs in dense clusters of 25–100 fruit bodies and is very abundant early in the fall season west of the Cascade Divide.

Microscopic characters. **Spore deposit** dingy cinnamon. **Spores** 5.5–7.5 x 3.7–4.5 (6) μ, smooth, in face view ovate to subelliptic, in profile obscurely inequilateral, with a well developed apical pore. **Basidia** 4-spored. **Pleurocystidia** none. **Cheilocystidia** 17–29 x 3.3–7 μ, fusoid-ventricose to nearly cylindric or some subcapitate. **Caulocystidia** scattered to fasciculate, resembling the cheilocystidia. **Epicutis of pileus** a gelatinous pellicle. **Clamps** present.

160 *Pholiota mutabilis*   *About natural size*

Field identification marks. (1) The bright bittersweet pink cap slowly fading to ochraceous in age and then developing black discolorations; (2) the bitter taste; (3) the bright yellow (to orange yellow) gills; (4) the base of the stalk dingy orange brown from handling.

Observations. Fruit bodies of this species are very difficult to dry. If overheated slightly or if they are too watersoaked when collected, they blacken in drying. The color of the spore deposit is a rusty brown; otherwise the species would be placed in *Naematoloma*.

Edibility. Not recommended.

When and where to find it. Solitary to gregarious on rotting conifer logs during the late summer and fall. It is not uncommon in the Pacific Northwest.

Microscopic characters. **Spores** 5–8 x 3.8–4.5 μ, smooth, rusty brown in deposit, ovate to elliptic in face view, obscurely inequilateral in profile. **Basidia** 4-spored. **Pleurocystidia** 35–60 x 8–14 μ, fusoid-ventricose to clavate-mucronate apex subacute, content as for chrysocystidia. **Cheilocystidia** 40–75 x 4–8 x 3–4.4 μ, slightly ventricose at the base, neck long and cylindric, apex obtuse. **Cuticle** of pileus a gelatinous pellicle of hyaline interwoven hyphae 2–5 μ wide. **Clamps** present.

161 *Pholiota astragalina*   *About natural size*

162 Pholiota fulvozonata

Field identification marks. (1) The zones of subferruginous to fulvous veil remnants on the stalk; (2) the habit of growing on recently burned areas (1–3 years after a fire); (3) gills pallid when young; (4) the cap not conspicuously scaly (some appressed scales present especially near the margin).

Observations. *Pholiota carbonaria* is very close to *P. fulvozonata* but has a truly ferruginous red veil, scales of this color are conspicuous in concentric rows on the cap, and its pleurocystidia measure 50–80 x 9–14 μ. However, there is a whole series of "variants" on burned ground with white, yellow, cinnamon, fulvous, russet, and red outer veils. In most of these there are also differences in other features also. Microscopic characters are needed for final identification.

Edibility. Not tested.

When and where to find it. On burned ground during cold wet weather in the fall; most material has been observed where brush piles have been burned during the course of road building.

Microscopic characters. **Spores** 6–7.5 x 4–4.5 μ, smooth, apical pore minute, shape in face view ovate to elliptic, in profile obscurely inequilateral. **Basidia** 4-spored. **Pleurocystidia** 33–46 x 9–16 μ, fusoid-ventricose to utriform, yellowish in KOH, thin-walled. **Cheilocystidia** and **caulocystidia** similar to pleurocystidia. **Pileus cuticle** a tangled mass of loosely arranged hyphae 2–3 μ wide, embedded in slime, more appressed near the surface than in the interior of the layer. **Veil hyphae** brown-walled in KOH, often with incrusted material on them and many of the cells short (to subglobose). **Clamps** present.

162 *Pholiota fulvozonata*　　　　　　　　　　　*About natural size*

163 *Pholiota subochracea*   *About two-thirds natural size*

Field identification marks. (1) The pale yellow sticky cap; (2) the yellow gills when young; (3) the stalk becoming rusty brown from the base up and lacking scales; (4) the odor and taste not distinctive.

Observations. For an accurate identification it is necessary to take into account the small spores and that the pleurocystidia are ± buried in the layer of basidia (not the elongated fusoid-ventricose type found in the subgenus *Flammuloides* in Smith and Hesler).

Edibility. Not tested and not recommended for testing.

When and where to find it. Solitary, gregarious or cespitose on decaying conifer logs during the fall season — often late fall. It occurs throughout the Pacific Northwest.

Microscopic characters. **Spore deposit** yellow brown. **Spores** 5–6 x 2.5–3 μ, smooth, in face view elliptic to oblong, in profile varying to slightly bean-shaped. **Basidia** 4-spored. **Pleurocystidia** present as chrysocystidia, 32–47 x 10–15 μ, clavate-mucronate to fusoid-ventricose. **Cheilocystidia** mostly similar to pleurocystidia. **Pileus cuticle** a layer of appressed gelatinous hyphae 2–4 μ wide. **Clamps** present.

Field identification marks. (1) The radiately arranged patches of appressed fibrils on the cap and their dull brown color; (2) the lack of a ring on the stalk (but a sheath of white fibrils present which breaks into squamules or patches); (3) habit of growing on slash (small branches rather than on stumps or logs).

Observations. As reported by Smith and Hesler (1968), this species appears to be important in the reduction of conifer slash following logging operations. It is a common and variable species, so for a final identification, check the microscopic features. Partly sterile basidiocarps with yellow gills have been found, an occasional collection exhibits a fragrant odor when fresh, and the color of the cap varies somewhat with the weather, being very dark vinaceous brown when wet and young.

Edibility. Not recommended.

When and where to find it. Solitary to scattered on small branches which have fallen to the ground and on slash, often abundant during cool wet fall weather; it occurs generally throughout the Northwest and Rocky Mountains.

Microscopic characters. **Spores** (5.5) 6–7.5 (8.5) x 3.5–4.5 μ, smooth, apical pore minute, shape in face view ovate to elliptic, in profile obscurely bean-shaped to obscurely inequilateral. **Basidia** 4-spored. **Pleurocystidia** 50–90 x (6) 9–18 μ, fusoid-ventricose with an obtuse apex, as revived in KOH some with thickened walls (1–2 μ thick). **Cheilocystidia** 36–55 x 8–12 μ, subfusoid to nearly clavate or fusoid-ventricose. **Pileus** with a pellicle of gelatinized hyphae about 2.5 μ wide, smooth, and thin-walled. **Clamps** present.

164 *Pholiota decorata*   *About natural size*

Strophariaceae

As the French mycologist L. Quélet realized years ago, the three genera *Psilocybe, Stropharia,* and *Naematoloma* (as *Hypholoma)* in reality constitute a single genus. It is a genus comparable to *Pholiota* in the sense of Smith and Hesler (1968). Since the genus has not been monographed for North America, the three older genera comprising it are here recognized.

Key to Genera

1. Stalk fleshy and with a ring . . . .(p. 217) *Stropharia*
1. Stalk lacking a ring 2
   2. Chrysocystidia present . . . .(p. 218) *Naematoloma*
   2. Chrysocystidia absent . . . .(p. 215) *Psilocybe*

Psilocybe

Key to Species

1. Cap merely moist, lacking greenish stains; thick-walled pleurocystidia present; stalk naked and polished; in swampy habitats . . . . (p. 215) *P. corneipes*
1. Cap tacky to the touch; often with greenish gray tints along the margin, pellicle separable . . .(p. 216) *P. pelliculosa*

Psilocybe corneipes 165

Field identification marks. (1) The shiny ochraceous orange to tawny cap; (2) the stalk strigose with dull tawny hairs at the base and the surface above this polished and shining; (3) context yellowish; (4) veil lacking; (5) spore deposit is brownish violaceous.

Observations. The thick-walled pleurocystidia and small spores are distinguishing microscopic features.

Edibility. Not known.

When and where to find it. Scattered to gregarious on wet mossy areas usually submerged in the spring. It is more abundant in the Pacific Northwest than anywhere else I have collected, but along the lower stretches of Upper Priest River in northern Idaho it is often so abundant as to be a prominent element in the agaric flora of swampy areas.

Microscopic characters. **Spores** dark brownish violaceous in deposit, 6–7 x 4–5 μ, nearly hyaline under the microscope, furnished with a hyaline pore, ellipsoid to slightly ventricose, smooth. **Basidia** 4-spored. **Pleurocystidia** abundant, 60–75 x 10–18 μ, fusoid-ventricose, thick-walled, apex occasionally incrusted. **Cheilocystidia** similar. **Cuticle** of pileus poorly differentiated. **Clamps** present.

165 *Psilocybe corneipes*　　　　　　　　　　　　*About natural size*

166　Psilocybe pelliculosa

Field identification marks.　　(1) The margin of the cap is straight at first, the cap conic to obtusely bell-shaped; (2) the context of the cap turns slightly bluish or greenish where injured; (3) the veil is absent to rudimentary; (4) the cap is

166 *Psilocybe pelliculosa*　　　　　　　　　　　　*About natural size*

sticky from a gelatinous pellicle; (5) the species grows on woody debris and humus rich in lignin.

Observations. The spore size is variable, in some collections it is 9.3–11 x 5–5.5 μ, and others 10.5–13 x 5.5–7 μ. This problem needs further study. It is a very ordinary appearing little mushroom.

Edibility. It is mildly poisonous; the action being on the nervous system to produce hallucinations. Other symptoms are also to be expected. I strongly urge people not to experiment with fungi of the blue-staining group in *Psilocybe.*

When and where to find it. Scattered to gregarious on wood and debris late in the fall season after cool wet weather in California, Idaho, Oregon, and Washington.

Microscopic characters. **Spores** 8–11 x 4.5–5.5 μ (some collections with spores 10.5–14 x 5.5–7 μ), dark yellow brown in KOH, with an apical pore, smooth, terete, ellipsoid to subovoid. **Basidia** 4-spored. **Pleurocystidia** none (or similar to cheilocystidia). **Cheilocystidia** 22–36 x 5–8 μ, hyaline, forming a sterile band, fusiform-lanceolate, apex acute to subacute. **Epicutis** of pileus a thick pellicle of appressed gelatinous hyphae 2–6 μ wide. **Clamps** present.

Stropharia

Stropharia kauffmanii   167

Field identification marks. (1) The grayish brown to tawny olive to dingy tawny cap which is densely innately squamulose to areolate (the latter over the disc); (2) the scaly condition of the stalk in the area below the ring; (3) the pallid gills at

167 *Stropharia kauffmanii*   *About one-half natural size*

first but becoming violaceous gray by maturity.

Observations. *Pholiota fulvosquamosa* is closest to this species, and the two furnish another example of closely related species being placed in different genera on the basis of the color of the spore depósit.

Edibility. I have no data on it.

When and where to find it. It fruits in the spring or fall in areas rich in decaying wood such as around brush piles that are in the last stages of decay, road fills containing much lignicolous debris, and similar situations. It is very likely more common and widespread than present records indicate.

Microscopic characters. **Spores** dark vinaceous brown in deposit (between "bone brown" and "army brown"), 6–7 x 4–4.5 μ, ellipsoid, smooth, apical germ pore very minute to absent. **Basidia** 4-spored. **Cheilocystidia** 50–60 x 10–13 μ, subcylindric to subfusoid, apex obtuse to acute, thin-walled, smooth, hyaline in KOH. **Cuticle** of pileus consisting of aggregations of fibrils with free tips (forming the scales of the pileus). **Clamps** present.

Naematoloma

Key to Species

1. Gills pallid at first; growing clustered on wood of conifers; cap reddish tan when fresh; taste mild . . . .
. . . .(p. 218) *N. capnoides*

1. Gills yellowish soon becoming olive to dull green; gregarious to clustered on wood of hardwood or conifers; taste typically bitter . . . .(p. 219) *N. fasciculare*

---

168   Naematoloma capnoides

Field identification marks. (1) The purplish gray gills when mature; (2) habit of growing in clusters on or around conifer logs and stumps; (3) lack of a ring on the stalk; (4) the reddish tan cap when mature.

Observations. *N. fasciculare* is common in the area also, and is distinguished by an olive to green shade in the gills. It is not recommended for the table. Both species occur on conifer wood mostly, and are among the first a beginning collector is likely to find during the fall season.

Edibility. *N. capnoides* is edible and a frequently collected "stump mushroom." *N. fasciculare* is NOT recommended — it is often bitter, and some people are poisoned by it.

When and where to find it. Both are common, occur in the fall on into the winter — especially in northern California.

Microscopic characters. **Spores** purplish in water mounts, yellow brown in KOH, 6–7 x 4–4.5 μ, smooth, apical pore present. **Pleurocystidia** present as chrysocystidia, 22–30 x 7–10 μ. **Pileus** with a thin nongelatinous pellicle. **Clamps** present.

168 *Naematoloma capnoides*   *About one-half natural size*

Naematoloma fasciculare   169

Field identification marks. (1) The purplish spore deposit; (2) the decidedly green to olive tint to the gills before these are colored by the spores; (3) growing in large clusters; (4) the typically bitter taste; (5) orange ochraceous cap becoming olive yellow over marginal area.

169 *Naematoloma fasciculare*   *About one-half natural size*

Observations. A mild tasting variant also occurs in the Northwest.

Edibility.   Not recommended. The typical variant has a persistently bitter taste. Some writers class it as poisonous and some as merely inedible. I have reports that the mild tasting variant is as good as *N. capnoides,* but would not endorse a general recommendation on this basis.

When and where to find it.   Common throughout the West on decaying wood of both conifers and hardwoods. It often fruits on into the winter if the weather is mild.

Microscopic characters.   **Spores** 6.5–8 x 3.5–4 μ, ellipsoid, pale dull yellow-brown mounted in KOH, purplish brown in water mounts, smooth, apex with a germ pore. **Basidia** 4-spored. **Pleurocystidia** (as chrysocystidia) abundant, 26–42 x 6–11 μ. **Cheilocystidia** 18–35 x 6–12 μ, obtusely fusoid-ventricose. **Gill trama** regular. **Cuticle** of pileus scarcely differentiated from trama. **Clamps** present.

Agaricaceae

Agaricus

The stalk is separable from the cap, a ring is present on the stalk unless the veil is very thin or granulose, the spores are some shade of chocolate brown, and the stalk can be separated from the cap with very little effort, and it breaks away cleanly.

Agaricus is a large and complicated genus and since few specialists know the species in any region of North America, the beginner does not need to apologize for his lack of information. Only a few of the species are treated here. One of the most difficult groups is the *A. arvensis* group consisting of white species with a "double" ring (see choice 5 of key). Mostly these stain yellow where bruised or where touched with a drop of potassium hydroxide (KOH), an aqueous solution of ± 5 percent can be used. Some of these are edible and some have caused mild cases of poisoning. *Agaricus crocodilinus* is the best for the table. It has a cap 15–30 cm broad which soon becomes coarsely scaly. It grows in pastures along the coast. It is recommended. *A. arvensis, A. sylvicola,* and *A. abruptibulba* are medium to large white nonscaly species known to be edible. These can easily be confused with *A. xanthoderma* and *A. albolutescens* which have been known to cause gastrointestinal upsets.

Key to Species

1.   Fruit body with tawny scales on cap, and patches and/or zones of ± tawny fibrils below ring on stalk; stalk 2–7 cm thick . . . .
. . . .(p. 226) *A. subrufescens* and (p. 222) *A. augustus*

1.   Not as above 2

Agaricus campestris   170
(Meadow Mushroom or Pink Bottom)

Field identification marks. (1) The white fibrillose surface of the cap; (2) the ring thin, consisting of a single layer of tissue, and often soon obliterated; (3) the pink gills of immature caps (chocolate brown when mature); (4) lack of a "cup" (or "volva") at base of stalk.

Observations. Two forms of the species are common in the West: the white, typical form and one with sparse to numerous brown fibrils on the cap.

Edibility. Edible, choice, and usually abundant. *Agaricus bisporus* is the name used for the commercially grown "species" in North America, It has 2-spored basidia; I have seen both the common white variant and the brown fibrillose one

170 *Agaricus campestris*   *Slightly less than natural size*

in the local markets. A species fruiting on hard-packed soil such as along roads, on playgrounds, etc., and has a double ring (a bandlike ring flaring at the top and bottom) is *Agaricus rodmani* (or in Europe known as *Agaricus edulis* or *Agaricus bitorquis).* It is also a much prized mushroom for the table.

When and where to find it. It grows scattered to gregarious in meadows, pastures, and grassy places generally. In the mountains it may fruit during late summer but fall is the best time; September and October are the best months to find it in Washington and most of Oregon, November through December are best in southern Oregon to California.

Microscopic characters. **Spores** 6–7.5 x 4.5–5 μ, ellipsoid to subovoid, dark chocolate brown in KOH. **Basidia** 4- or (rarely, some) 2-spored. **Gill trama** subparallel to interwoven, subhymenium cellular and well developed. **Cuticle** of pileus scarcely distinguishable from the context, hyaline in KOH.

171 Agaricus augustus (The Prince)

Field identification marks. (1) The large size (caps up to 15 inches wide); (2) the crust brown to darker appressed fibrillose scales; (3) a sheath (usually broken into zones) of ± crust brown fibrils below the ring; (4) bruised areas tend to stain yellow; (5) the ring is double to the extent of having patches of tissue on the underside.

Observations. The large spores distinguish this *Agaricus* from others in the genus which have ± tawny brown fibrils on the cap.

Edibility. Choice. The chief regret I hear from collectors is that it is seldom found in quantity.

When and where to find it. Solitary to scattered along roads, on waste ground, piles of decaying debris, and ant hills in the spring and fall. It is more abundant along the coast than elsewhere but is not a "common" species anywhere in the western area.

Microscopic characters. **Spores** 8–11 x 5–6.5 μ, ellipsoid to subovoid, smooth, chocolate brown in KOH when mature. **Basidia** 4-spored. **Pleurocystidia** none. **Cheilocystidia** of two types: saccate and 8–15 x 7–10 μ, and fusoid-ventricose measuring 22–34 x 8–12 μ. **Pileus cuticle** of radial hyphae 5–10 μ wide and yellowish in KOH. **Clamps** absent.

172 Agaricus subrutilescens

Field identification marks. (1) The dark lilac brown to vinaceous brown fibrils over the central part of the cap; (2) the tendency for these fibrils to be arranged in patches near the cap margin; (3) below the ring the stalk is covered by a soft sheath of white fibrils or is at times tomentose from soft hairs projecting from the sheath; (4) the ring often appears to be single but usually some patches occur on the underside or along the margin.

171 *Agaricus augustus*   *About two-thirds natural size*

172 *Agaricus subrutilescens*   *Slightly less than natural size*

Edibility. Edible and one of the best flavored mushrooms I have tried, but some people, myself included, cannot tolerate it. Each individual must try it for himself, and should not over indulge.

When and where to find it. Gregarious to scattered under conifers, especially with Sitka spruce, along the coast during the fall months, but may at times be common inland.

Microscopic characters. **Spores** 5–6 x 3–3.5 μ, ellipsoid, smooth, dark chocolate brown in KOH. **Pleurocystidia** none. **Cheilocystidia** 10–20 x 8–15 μ, saccate to clavate. **Cuticle** of pileus of radial hyphae 4–9 μ wide, their walls violaceous brown. **Clamps** none.

---

173 Agaricus hondensis

Field identification marks. (1) Cap smooth or with patches of appressed fibrils and these patches darkening and becoming more conspicuous in age; (2) cap fibrils darkening to grayish vinaceous and finally vinaceous brown; (3) stalk having an abrupt basal bulb; (4) stalk naked below the ring (no layer of fibrils as in *A. subrutilescens*); (5) underside of ring with soft patches of grayish vinaceous fibrils; (6) flesh usually staining yellowish then pinkish and finally dingy vinaceous.

Observations. Previously I recognized two species: *A. hondensis* and *A. silvaticus,* the latter having the more scaly cap. However, in recent times an apparently different species has had the latter name applied to it. At present I have decided to use an American name for our species. *Agaricus bivelatoides, A. hillii, A. mcmurphyi,* and *A. glaber* in addition to *A. hondensis* are available insofar as I think they all apply to one and the same species. Since the fungus I previously

173 *Agaricus hondensis*   *About two-thirds natural si*

identified as *Agaricus silvaticus* also falls in this group, I am illustrating the latter here under the name *A. hondensis*.

Edibility. Not recommended. The species concepts have been confusing, to say the least, and it appears that some populations are poisonous at least to some people.

When and where to find it. Scattered to gregarious in mixed conifer-hardwood forests in the fall along the coast from British Columbia to California, ordinarily not abundant.

Microscopic characters. **Spores** 4.5–6 x (3)3.5–4 μ, ellipsoid. **Basidia** 4-spored. **Pleurocystidia** none. **Cheilocystidia** basidiumlike. **Pileus cuticle** of hyphae 3.5–8 μ wide, walls pale vinaceous brown in KOH. **Clamps** absent.

Agaricus placomyces (The Flat-Capped Agaricus) 174

Field identification marks. (1) The yellow to brown drops on the underside of the unbroken veil which become darker brown usually by the time the veil breaks; (2) and the thin coating of grayish fibrils forming minute appressed squamules over the cap.

Observations. This species has been frequently reported for our western area, but it now appears that other species were involved among them, for instance *Agaricus meleagris*. *A. meleagris* does not have the droplets on the underside of the ring. All previous authors including myself have contributed to this confusion. *A. meleagris* is the common species.

Edibility. Edible as far as my information goes, but the statements in the literature apply mostly to *A. meleagris* — which is poisonous to some people.

When and where to find it. Scattered to gregarious on rich

174 *Agaricus placomyces*   *Slightly less than natural size*

humus in low hardwood forests, Great Lakes area, and eastward during the late summer and early fall. Rare during most seasons. I have seen no authentic specimens from west of the Great Plains.

Microscopic characters. **Spores** dark vinaceous brown ("bone brown") in deposit, 4–5.5 x 3.5–4 μ, ellipsoid, smooth. **Basidia** 4-spored.

---

175   Agaricus sylvicola

Field identification marks.   (1) The fruit body is white except for the gills; (2) KOH dropped on the cap stains it yellow; (3) the gills are grayish pink but become brighter before becoming chocolate brown; (4) the ring is double and the underlayer separates as shown in the photograph; (5) the stalk is flattened at the base, a flaring bulb may or may not be present.

Observations.   There are many white species with a double ring, but as yet we have no authoritative work on the North American representatives.

Edibility.   A. arvensis and A. sylvicola are edible but some closely related species are not, hence I do not recommend any in the group.

When and where to find it.   On humus in both conifer and hardwood forests. I have not been able to ascertain any particular preference as to habitat for *A. sylvicola*. It fruits during the summer and fall.

Microscopic characters.   **Spores** 5–6.5 x 4–4.5 μ, subovoid to ellipsoid. **Basidia** 4-spored. **Pleurocystidia** none. **Cheilocystidia** abundant, 15–28 x 8–15 μ, clavate to fusoid-ventricose. **Pileus** with a poorly defined cuticle, the hyphae 5–8 μ wide. **Clamps** absent.

---

176   Agaricus subrufescens

Field identification marks.   (1) The tawny to pale tawny cap; (2) the sheath of the stalk breaking up into squamules or patches; (3) the conspicuous floccose patches on the underside of the ring.

Observations.   The above features identify a group of species differentiated by spore size. *Agaricus perrarus* is common in the Sitka spruce zone along the Oregon and California coast. It, like *A. augustus,* has large spores, 8–10(12) x 4.5–6 μ, but its colors are more ochraceous.

Edibility.   Edible. Both species are popular on the West Coast.

75 *Agaricus sylvicola*   *About natural size*

76 *Agaricus subrufescens*   *About one-third natural size*

When and where to find it. Solitary to gregarious on rich humus or around piles of organic debris in the fall. I have always regarded these species as being worth some study by gardeners to help reduce compost piles.

Microscopic characters. **Spores** 6–7.5 x 4–5 μ, ellipsoid, dark chocolate brown in KOH, smooth, no germ pore evident at ordinary magnifications. **Basidia** 4-spored.

Coprinaceae

This family is distinguished primarily on the structure of the skin of the cap which is composed of inflated hyphal cells, dark colored spore deposit, and the presence of an apical germ pore in the spores of most species. It is difficult to key out on field characters, but one soon learns to recognize it by the very fragile consistency of the fruit bodies of most species. There are over 500 kinds known for North America alone.

Key to Genera

1. Gills liquefying (undergoing autodigestion) (p. 228) *Coprinus*
1. Gills not undergoing autodigestion . . . . . . (p. 232) *Psathyrella*

Coprinus (The Inky Caps)

The major characteristics of this genus are: Spore deposit black to blackish brown; gills undergoing auto-digestion (liquefying) after spore discharge; caps variable in size and with a granular to fibrillose veil or none; growing on soil, manure, rubbish heaps, or plant remains generally.

Key to Species

1. Fruit bodies crust brown to pale tan, at first with ± inconspicuous granular veil remnants; densely gregarious to clustered . . . . (p. 229) *C. micaceus*
1. Not as above 2
   2. Cap ± oval and 4–10(15) cm tall; smooth on apex, fibrillose lacerate to scaly down the side . . (p. 230) *C. comatus*
   2. Not as above 3
3. In large clusters, cap glabrous or with appressed scales; with a slight brownish zone at base of the stalk; spores smooth . . . . (p. 231) *C. atramentarius*
3. Solitary or in small clusters (2–3); cap at first with remains of a thin white veil; no colored zone at base of stalk; spores warty . . . . (p. 232) *C. insignis*

Coprinus micaceus

177

(Mica Cap)

Field identification marks.   (1) The tawny to pale crust brown cap when young appearing as if sprinkled with fine sugar; (2) the clustered growth pattern; (3) the blackening and deliquescing gills.

Observations.   In the western area the common form of the species occurs closely gregarious rather than in clusters, but in open areas one finds both types often intermingled. In Europe *C. truncorum* and *C. micaceus* have been recognized as closely related species.

Edibility.   Edible. I get conflicting reports on this species: some rate it highly and others complain that it is very mild. Be sure to cook this mushroom soon after collecting it because it deliquesces very rapidly otherwise. Caps in which the gills have started to darken will deliquesce in an hour or so if placed in a plastic bag and subjected to room temperature.

When and where to find it.   Around old stumps, buried wood, and rotting logs of hardwoods. It fruits both during the spring and fall season and is one of our most common mushrooms throughout the area where hardwood trees (broad-leaved trees) grow. Numerous crops can be obtained from a single stump.

Microscopic characters.   **Spores** 7–9 x 4–5 μ. **Pleurocystidia** 40–85 x 10–35 μ. **Veil particles** consisting of ± globose cells 20–50 μ in diameter, some with pedicels.

177 *Coprinus micaceus*   *Slightly less than natural size*

178 Coprinus comatus
(Shaggy-Mane)

Field identification marks. (1) A movable ring is present on the stalk (this frequently becomes broken and falls off); (2) the elongate scaly cap with smooth apex (resembling a skull cap); (3) gills white, then reddening, then black and deliquescing.

Observations. This is one of the most common and distinctive species of gilled mushrooms, easily identified at sight.

Edibility. Edible and choice. Button stages in which the gills have not changed color are best. DO NOT keep fresh specimens in the refrigerator overnight as they are likely to deliquesce!

When and where to find it. Gregarious on lawns, in waste areas, along roads, playing fields where soil is packed, etc. Common throughout the area during late summer or fall, rarely in the spring. One should take caution about collecting for the table along highways, especially if the right of way has been sprayed. Some cases of digestive disturbances have probably been caused from sprays absorbed by the mushroom.

Microscopic characters. **Spores** 11–14(15) x 6.5–8(8.5) μ, smooth, with a hyaline subapical germ pore. **Clamps** present.

178 *Coprinus comatus*   *About one-third natural size*

(Inky Cap)

Field identification marks.   (1) A basal zone of brown fibrils on the stalk; (2) generally gray to gray brown cap which may or may not be scaly around the central area; (3) growing in dense clusters. As an inky cap in the broad sense of the word, the gills undergo autodigestion (deliquesce) after the spores have been discharged.

Observations.   Both a scaly and a glabrous variant are common, but the scaly one is to be regarded as typical since Fries mentioned the scaly disc in the validating description in 1821.

Edibility.   Edible and used widely but an occasional person experiences a peculiar type of intoxication if he or she drinks an alcoholic beverage within a few hours after eating this species. This reaction is well known as the alcohol-disulfiram syndrome but disulfiram itself has not been detected in *C. atramentarius.* Apparently the raw mushroom does not produce the effect.

When and where to find it.   It "feeds" on buried wood and fruits both in the spring and fall throughout the area. Most people find it right in their own yard. When building his house and making his lawn, one man had a sharp depression in his property in which he dumped all the waste wood and also some stumps and dead roots which he desired to get rid of, and then covered this with dirt and planted grass over it. Two years later he had a "Coprinus explosion" of hundreds of large clusters of this species.

Microscopic characters.   **Spores** (7)8–10 x 4.2–5 μ. **Brachybasidioles** 11–15 x 9–12 μ. **Pleurocystidia** 150 x 30 μ. **Clamps** present.

179 *Coprinus atramentarius*   *Slightly less than natural size*

180   Coprinus insignis

Field identification marks.   (1) A thin silky fibrillose veil evident on most immature caps; (2) the rimose-striate margin of the cap near maturity; (3) occurrence around the base of hardwood trees, especially maple; (4) caps 2–4.5(5) cm high; (5) the warty-roughened spores (not a field character but essential to correct identification of the fungus).

Observations.   The appearance of this species strongly suggests *C. atramentarius* but the latter does not have a thin white veil when young and its spores are smooth.

Edibility.   POISONOUS. Apparently, this is the most poisonous species in the genus, at least the material collected in Europe indicates this. I have no data on its edibility, but report it here as poisonous more as a caution to those eating *C. atramentarius.*

When and where to find it.   Solitary or in small clusters around rotting hardwoods, especially maple, during the summer and fall. I consider it likely to occur in the West because of the occurrence of *Acer* in the area, but have not yet found it.

Microscopic characters.   **Spores** 10–13 x 6–7.5 μ, verrucose, with a distinct apical pore. **Clamps** present.

Psathyrella

Key to Species

181   Psathyrella longistriata

Field identification marks.   (1) The fragile consistency of the cap and its gray brown to cinnamon color when moist (it is much paler faded); (2) the prominent ring which is conspicuously striate on the upper surface; (3) the fragile whitish stalk usually has scattered patches of fibrils below the ring.

Observations.   To the above one might add that along the Pacific Coast it is likely to be the first annulate *Psathyrella* one encounters. The color of the moist cap varies considerably, and a few variants distinguished largely on microscopic characters are known but are apparently rare.

Edibility.   Not recommended. It has little substance, and reliable data are lacking.

180 *Coprinus insignis*   *Slightly less than natural size*

181 *Psathyrella longistriata*   *Slightly less than natural size*

When and where to find it. It appears during the late summer and fall after heavy rains, solitary, gregarious, or subcespitose in groups of 2–3, on duff under alder and conifers mixed.

Microscopic characters. **Spore deposit** vinaceous brown to purplish brown. **Spores** 7–9 x 4–5 μ, in face view elliptic to ovate, in profile obscurely inequilateral, apical pore inconspicuous. **Pleurocystidia** 40–60 (72) x 10–17 μ, fusoid-ventricose, apex obtuse to subacute, thin-walled. **Cuticle** of pileus of vesiculose and pear-shaped cells mixed. **Clamps** present.

182   Psathyrella carbonicola

Field identification marks. (1) The conspicuous remains of a white fibrillose outer veil over the cap; (2) the reddish brown ground color when moist (beneath the veil); (3) habitat on burned areas; (4) the stalk darkening at the base and being typically squamulose over the lower part from remains of the veil.

Observations. The veil remnants on the cap are superficial and easily removed by hard rain or other mechanical intervention.

Edibility. Not to be regarded as an esculent because of lack of substance. I have no data on whether it is poisonous or not.

When and where to find it. Scattered, gregarious, or cespitose on burned areas during the summer and fall; often abundant. The *Pholiota* most frequently found growing with it is *P. highlandensis.*

Microscopic characters. **Spores** 6–7 x 3–3.5 μ, smooth, ellipsoid to somewhat ovoid, apical pore minute. **Basidia** 4-spored.

182 *Psathyrella carbonicola*

About natural size

Pleurocystidia and **cheilocystidia** abundant and similar, 34–46(50) x 9–12 μ, hyaline, fusoid-ventricose, apex acute, thin-walled. **Clamps** present.

Russulaceae

This family features heteromerous trama of the cap and stalk, and spores with amyloid ornamentation. In the North American species there is no appreciable development of a veil, and the stalk in *Russula* at least is very fragile. In *Lactarius* a latex is present, and the fruit bodies are coarse and short-stalked. The family represents the agaricoid end of a line of gasteromycetous genera: *Macowanites, Elasmomyces, Arcangelliella, Martellia, Gymnomyces,* etc.

Key to Genera

1. Latex absent (be sure to test young fruit bodies) . . . .
. . . . (p. 235) *Russula*
1. Latex present (cut apex of stalk slightly with a sharp instrument) . . . . (p. 237) *Lactarius*

Russula

Key to Species

1. Cap olivaceous becoming splashed dark red or finally entirely red; pellicle not cleanly separable from flesh
. . . . (p. 235) *R. olivacea*
1. Cap olivaceous, not becoming red; pellicle ± separable from flesh; often found under aspen . . . .
. . . . (p. 236) *R. aeruginea*

Russula olivacea 183

Field identification marks. (1) The large size (caps up to 35 cm broad occur); (2) pileus olive-colored or olive mixed with vinaceous red in various degrees; (3) dingy yellowish gills; (4) thick stalk (2.5–6 cm), often dingy rose-tinted; (5) cap surface soon dry and unpolished; (6) the mild taste; (7) becoming eosine red where touched with phenol.

Observations. This is our largest *Russula.* The specimens photographed justify the species epithet but one should not expect to find the cap so evenly colored. It is usually laced with vinaceous red in varying amounts and degrees, but seldom entirely red. The color of the stalk varies in intensity also.

Edibility. Collectors have reported *R. olivacea* as an edible species, but I have no data of my own on it. One large button

183 *Russula olivacea*   *About one-half natural size*

would make a meal for two people. When experimenting with it observe the usual precautions.

When and where to find it. Solitary to gregarious in seepage areas and along streams in old-growth mossy conifer forests in areas of high precipitation. It often occurs under Devil's Club. It can be collected regularly in the Priest Lake district of Idaho and in the Olympic Mountains of Washington. It fruits during the summer and fall seasons.

Microscopic characters. **Spore deposit** yellow. **Spores** 8–11 x 7–9 (10) μ, subglobose, ornamented with amyloid warts 0.5–1 μ high, plage area amyloid. **Basidia** 4-spored. **Pleurocystidia** 60–90 x 8–15 μ (72–100 x 8.5–13 [16] μ, Romagnesi), versiform: fusoid, subcylindric-pointed, or capitate. **Cuticle** of pileus a lax turf, the end cells somewhat cystidioid, the other cells of the filaments variously slightly inflated to subcylindric.

---

184   Russula aeruginea

Field identification marks. (1) The cap is glabrous, thinly slimy when wet, and dull green; (2) the taste is mild or nearly so; (3) the gills are soon spotted brownish where damaged by insects, and the base of the stalk is usually with cinnamon buff (pale tan) discolorations; (4) the pellicle of the cap is separable about half way to the disc.

Observations. The pale yellow spore deposit should be added to the above features for an accurate identification.

Edibility. Edible. I have not tried it.

184 *Russula aeruginea*   *About two-thirds natural size*

When and where to find it.   In our western region I have
seen it most abundant in lodgepole pine-aspen mixtures from
mid-July to early fall, especially in central Idaho, but it is
widely distributed in the western area.

Microscopic characters.   **Spores** near cream color in a
deposit, 6–9 (10) x 5–7 μ, broadly ellipsoid, ornamentation
of warts and some fine radiating lines but not reticulate,
prominences up to 0.5 μ high. **Basidia** 4-spored. **Cystidia**
45–70 (80) x 7–12 μ, cylindric to subfusiform, apex versiform:
obtuse, capitate or with a protrusion. **Cuticle** of pileus a turf
of branched hyphae and some pileocystidia also present.
Clamps absent.

Lactarius

This genus, distinguished from *Russula* by the presence of latex in the fruit body, is a popular genus with
mushroom collectors, and is very conspicuous in the
fall flora.

Key to Species

1.   Latex dark red to carrot color 2
1.   Latex not as above 3
   2.   Latex dark muddy red when first exposed . . . .:
     . . . . (p. 238) *L. sanguifluus*
   2.   Latex ± orange at first . . . . (p. 238) *L. deliciosus*
3.   Cap glabrous, yellow; latex staining gills and flesh
    violaceous . . . . (p. 240) *L. aspideus*
3.   Not as above 4

4. Cap yellow; stalk with large spots; latex white and changing to yellow or staining yellow; cap margin hairy
.............................(p. 241) *L. scrobiculatus*

4. Not as above 5

5. Cap vinaceous brown moist; latex waterlike; cap margin naked; odor fragrant ....(p. 242) *L. aquifluus*

5. Cap pinkish to dull pale rose over center, margin coarsely hairy; latex milk white; odor not distinctive
......(p. 243) *L. torminosus* var. *nordmanensis* comb. nov.*

185   Lactarius sanguifluus

Field identification marks. (1) The dark blood red latex which shows at the apex of the stalk if the latter is cut with a sharp razor; (2) the tendency for bruised areas to stain green; (3) the dull purplish red gills in mass as one views the gill layer.

Observations. If the latex or "milk" is carrot color, one most likely has one of the variants of *L. deliciosus,* a variable species in the West but edible, though some of the variants are more desirable for the table than others. The green staining is also a feature of *L. deliciosus.*

Edibility. Edible, and preferred by many to *L. deliciosus* which, according to the name, should be delicious but is not always so.

ﬀen and where to find it. Most abundant between the Cascades and the ocean under pine and also in stands of second growth Douglas fir, late summer and fall. I have seen it most prolific during warm wet seasons in southern Oregon, and on Hoodoo Mountain near Priest River, Idaho.

Microscopic characters. **Spores** 7.5–9.5 x 5.5–7 μ, broadly ellipsoid, finely reticulate. **Macrocystidia** (32) 40–60 (80) x 4–8 μ. **Pileus cuticle** a thick gelatinous pellicle.

186   Lactarius deliciosus

Field identification marks. (1) The cap is a grayish to ± bright carrot color; (2) latex is ± carrot orange when first exposed, but injured places stain green; (3) flesh of cap in young fruit bodies lacking blue tints.

Observations. A number of variations are known from our western states, but all are edible.

Edibility. Edible and choice, but apparently populations differ somewhat in their taste so that one should be prepared to try it more than once. Slow cooking is recommended.

L. nordmanensis Smith, Brittonia 12:308. 1960.

85 *Lactarius sanguifluus*   *About one-half natural size*

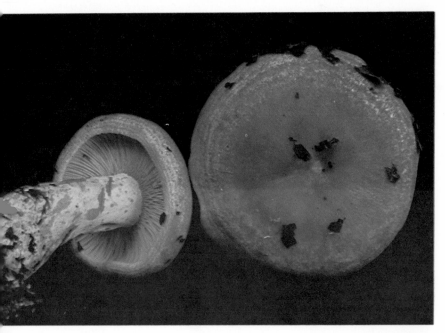

86 *Lactarius deliciosus*   *About two-thirds natural size*

When and where to find it. In wet conifer forests often under
Devil's Club and along streams, summer and fall; common.
The fruit bodies persist for a considerable time so that many
are full of insect larvae when found.

Microscopic characters. **Spore deposit** yellowish. **Spores**
8–10 x 6–8 μ, broadly ellipsoid to subglobose, coarsely netted
(mount in Melzer's). **Basidia** 4-spored. **Pleurocystidia:** macrocystidia 42–60 x 5–8 μ, fusoid to aciculate; pseudocystidia
rare, filamentose. **Cheilocystidia** 42–60 x 5–8 μ, ± resembling macrocystidia. **Pileus cuticle** an ixocutis. **Clamps** none.

Field identification marks. (1) Latex staining the gills violaceous where wounded; (2) cap yellow and lacking a strigose margin when young; (3) taste both acrid and bitter; (4) stalk sticky when young and moist.

Observations. If the cap margin is strongly strigose at first and the collection has all the other characters given above, one has a specimen of *L. representaneus* which is also common at times in our western area. It usually occurs under spruce. *L. aspideus* occurs in the oak forests of southern Oregon and northern California.

Edibility. Both the above should be regarded as dangerous, enough so as to discourage any experimentation.

When and where to find it. Solitary to gregarious on humus, under oak and madrone in the fall often in rather dry situations. It is rare in the western area but I have seen it near Takilma, Oregon. *L. repraesentaneus* is common.

Microscopic characters. **Spore deposit** white. **Spores** 7–9 x 6–7 μ, broadly ellipsoid to subglobose, with scattered very small warts, a few bands of amyloid material and lighter lines leading from them but not forming a reticulum. **Basidia** 4-spored. **Pleurocystidia** abundant, narrowly fusoid to lanceolate, 50–70 x 6–10 μ, content refringent. **Cheilocystidia** similar to pleurocystidia. **Cuticle** of pileus a layer of interwoven hyphae giving rise on surface to a dense turf of narrow yellowish filaments. **Clamps** absent.

187 *Lactarius aspideus*   About two-thirds natural s

Field identification marks. (1) The white latex soon changing to sulphur yellow; (2) the acrid taste with a burning aftertaste; (3) the yellow, sticky cap with a bearded margin; (4) the stalk is scrobiculate-spotted.

Observations. In addition to the above features, the spore deposit is white. There are a number of yellow *Lactarii* with latex changing to yellow, see Hesler and Smith (1960).

Edibility. Not recommended.

When and where to find it. Scattered to gregarious in the conifer forests of our western mountains. It is particularly abundant in the Olympic National Park late in the summer and early fall, and in the Priest Lake district of Idaho.

Microscopic characters. **Spores** white in deposit, 7–9 x 6–7 μ, ellipsoid to subglobose, ornamentation in the form of broken lines and isolated warts, but at times fused to a broken reticulum, with fine lines extending out from the main ridges, the prominences 0.5–1 μ high. **Basidia** 4-spored. **Pleurocystidia** 60–100 x 6–11 μ, cylindric to an aciculate apex, filamentous or narrowly clavate (the latter often embedded in the hymenium), content refractive in KOH. **Epicutis** of pileus a very thick layer of gelatinous hyphae 2–3 μ in diam. and nearly hyaline in KOH.

188 *Lactarius scrobiculatus*   *About one-half natural size*

189 *Lactarius aquifluus*

About two-thirds natural s.

189   Lactarius aquifluus

Field identification marks.   (1) The pronounced fragrant odor which is persistent for years in the dried specimens; (2) the watery latex; (3) the vinaceous brown to vinaceous cinnamon colors; (4) the context not staining when cut.

Observations. *L. helvus,* as described by Fries, has an acrid taste and white (not watery) latex. *L. aquifluus* Peck was described as having a waterlike latex but has long passed under the name *L. helvus* in North America.

Edibility.   NOT recommended. It is considered poisonous by some European authors, and edible but worthless by some American writers. One can find details of the poisonous nature of *L. helvus* in Pilát and Ušák, p. 65.

When and where to find it.   Scattered to gregarious under 2-needle pines in wet areas such as seepage areas in stands of lodgepole, but is not limited to this habitat. It is abundant when it fruits, and it fruits nearly every fall season. I regard it as one of the mushroom weeds of conifer country. Both in Switzerland and in North America I have found it to favor areas near dried up woodland pools with pine in the background.

Microscopic characters.   **Spore deposit** white. **Spores** 7–9 x 5.5–7 μ, broadly ellipsoid, verrucose-reticulate. **Basidia** 4-spored. **Pleurocystidia** 40–60 x 9–13 μ, rare, fusoid, embedded or only the tips projecting. **Cheilocystidia** mostly clavate to cylindric, apex obtuse. **Cuticle** of pileus a tangled mass of hyphae in a lax turf, the tips often grouped into fascicles.

Field identification marks. (1) The very hairy-strigose margin of the cap; (2) the pinkish tone over the central part; (3) the very acrid ("hot") taste of the raw flesh; (4) the yellowish spore deposit; (5) the white latex staining the gills yellow.

Observations. A very similar *Lactarius* also occurs in the West, but in it the latex does not stain the gills yellow on exposure. It is *L. torminosus* var. *torminosus*.

Edibility. For var. *torminosus,* edible if properly prepared but I do not recommend it, since there have been cases of poisonings. We have no records on var. *nordmanensis*.

When and where to find it. It forms mycorrhiza with *Betula* species (birches) and is to be expected where this genus occurs. It is very abundant at times in the Priest Lake district of Idaho during late summer and early fall. It also occurs in lawns where birch has been planted for ornamental purposes.

Microscopic characters. **Spores** 7.5–9(10) x 6–7.5 μ, ellipsoid, ornamented with a broken network of lines and some isolated warts. **Macrocystidia** 45–52 x 6–8 μ, fusoid-ventricose, acute at apex. **Pileus cuticle** a poorly defined ixocutis.

190 *Lactarius torminosus var. nordmanensis*   *Slightly less than natural size*

Puffballs and Related Fungi

Since the treatment here is mainly of individual genera, no diagnosis is given for each. The following key emphasizes the generic characters. Collectors not equipped with a microscope should simply compare their collections with the illustrations.

Key to Genera

1. Fruit body resembling an unexpanded mushroom (in a longitudinal section a cap, gill cavity [gleba] and stalk visible), spores not forcibly discharged from basidia 2
1. Not as above 3
   2. Appearance of an unexpanded *Russula;* gleba often exposed over area around the stalk; stalk short and exceedingly fragile . . . . (p. 252) *Macowanites*
   2. Appearance more resembling an unexpanded *Agaricus* or *Lepiota;* gleba not exposed around stalk or finally only slightly so; stalk ± fibrous . . . . . . (p. 250) *Endoptychum*
3. Fruiting body expanding to ± star-shaped . . . . *Geastrum* and (p. 254) *Astreus* (only the latter illustrated here)
3. Not as above 4
   4. Fruit body in form of a nest or vase, containing small pill-like objects (peridioles) . . . . (p. 256) *Nidula*
   4. Not as above 5
5. Small puffballs lacking a true stalk and opening by an apical pore; sterile base (area below gleba) typically chambered when present . . . . (p. 252) *Lycoperdon*
5. Not as above 6
   6. Aspect of a *Lycoperdon* but apex of fruit body rupturing and falling away, the sides remaining intact as a bowl-like structure containing the powdery gleba . . . . (p. 249) *Vascellum*
   6. Not as above, wall of fruit body gradually breaking up and falling away to expose the ± powdery mature gleba 7
7. Capillitial threads smooth and ± lacking side branches . . . . (p. 245) *Calvatia*
7. Capillitium of threads with short side branches (± thorn- or antler-like) . . . . (p. 248) *Calbovista*

Calvatia

The characteristics of this familiar puffball genus are: fruit bodies medium to large; gleba chambered and chambers lined by hymenium; gleba cottony to powdery at maturity; fruit body with the walls fracturing into pieces and these falling away exposing the gleba. *Calbovista subsculpta* is a typical *Calvatia* except for the character of the capillitium.

Key to Species

1. Fruit body 20–60 cm broad; growing in sagebrush areas; warts large and pallid to yellow brown . . . . (p. 246) *Calvatia booniana*
1. Fruit body smaller and growing in or near conifer forests in the mountains 2
    2. Fruit body 2–10 cm broad, warts of surface grayish to fuscous . . . . (p. 247) *Calvatia subcretacea*
    2. Fruit body with large areolate warts eventually falling away (much as in *Calvatia booniana*) . . . . (p. 248) *Calbovista subsculpta*

191   Calvatia booniana

Field identification marks.   (1) Fruit body flattened-globose and up to 60 cm wide (about 2 ft.); (2) the large flat scales which usually form by maturity; (3) habitat in sagebrush country.

Observations.   The western United States has more than its share of large odd puffballs. All are edible if white clear through and homogeneous in texture. It is curious that one of the largest prefers semi-desert habitats.

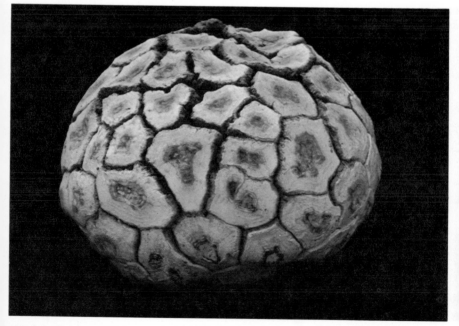

191 *Calvatia booniana*   *About one-fifth natural size*

Edibility.   Edible. But be sure the interior has not started to yellow.

When and where to find it.   It occurs in the sagebrush areas of the West after heavy summer rains, but the limits of its range have not been established. The collections in the herbarium at the University of Michigan have come mainly from the Owyhee Mountains of southwest Idaho.

Microscopic characters.   **Capillitial threads** 3.3–8.8 μ wide, branched, branching Y-shaped, tips of branches bullet-shaped to subacute, rarely thornlike; threads septate and readily breaking at the septa, walls with occasional pits, the pits often not extending clear through the wall. **Spores** 4–6 (6.5) x 3.5–5.5 μ, globose to broadly elliptic, very finely punctate, pedicel usually evident.

Calvatia subcretacea   192

Field identification marks.   (1) The dark colored warts over the peridium (wall of spore sac); (2) habit of growing solitary under conifers in mountain forests (spruce and fir); (3) the small to medium size of the fruit bodies for the genus.

Observations.   The warts break away individually at maturity. The slitlike pits in the capillitium and almost smooth spores which are cinnamon brown to olive brown in KOH are also important characters.

Edibility.   Probably edible, but one is not likely to find enough for a meal.

When and where to find it.   It can be collected frequently in the mountains around McCall, Idaho, but even there it is found solitary or only a few in a large area. It is a summer to early fall species of the spruce-fir zone.

192 *Calvatia subcretacea*   *About one-half natural size*

Microscopic characters. **Capillitium** of threads becoming attenuated to subacute at apex, (3)5–12(22) μ wide, aseptate or rarely septate, breaking readily, dark olive brown in KOH when mature; pits slitlike. **Spores** (3.3)4–6.5 μ, globose to subglobose, nearly smooth, the ornamentation ± 0.25 μ high in heavily ornamented individuals, cinnamon brown to dark olive brown in KOH.

Calbovista

193 Calbovista subsculpta

Field identification marks. See observations.

Observations. There is no way to distinguish this species from members of *Calvatia* without using microscopic characters. The flat scales are a feature of some collections but in others they may be somewhat pointed, and often show parallel horizontal markings as in *Calvatia sculpta*. A sterile base is present and becomes purplish in weathering.

Edibility. Edible when white throughout the interior.

When and where to find it. Solitary to gregarious in mountain conifer forests, especially near their edges, or in grassy openings, from April to August. It is one of the more common mountain puffballs in the West.

Microscopic characters. **Capillitium** free, consisting of short discrete units with abundant antlerlike branching, much entangled; secondary branches bluntly pointed, not varying much in width from main branch; threads 5–10 μ wide; wall up to 2.5 μ thick but thinner toward the tips, neither septate nor pitted, ochraceous yellow in KOH.

193 *Calbovista subsculpta*   *About one-fourth natural size*

Vascellum lloydianum   194

Field identification marks. (1) Aspect of a small "puffball" (2–6 cm wide) which opens by fracturing over the apex with the sides usually left intact so that the old ripe fruit bodies resemble ± a small bowl; (2) the gleba is powdery; (3) the young fruit bodies are white clear through and the outside is ± covered with warts and spines which slowly fall off.

Observations. The genus *Vascellum* is intermediate between *Calvatia* and *Lycoperdon*. In fact most species of *Vascellum* were formerly placed in *Lycoperdon*.

Edibility. Edible when the gleba is white.

When and where to find it. Common on lawns and grassy places generally west of the Cascade Mountains, usually fruiting late into the fall.

Microscopic characters. **Spores** 3.3–4.5(5) μ, globose, nearly smooth (under oil immersion surface minutely warty), ± argillaceous in KOH, tawny in Melzer's, often with a stub of a pedicel. **Eucapillitium** scant, threads 3–5 μ wide, walls brown in KOH, unpitted, tapered to acute apices; paracapillitium of hyaline ± thin-walled threads 2.5–6 μ wide. **Exoperidium** of hyaline pseudoparenchyma giving rise to heaps and cones of hyphae and hyphal cells. **Endoperidium** of thick-walled tubular hyphae, 2–7 μ thick, with ± colored walls and lumen almost obscured. No mycosclerids observed.

194 *Vascellum lloydianum*   *About natural size*

Key to Species

1. Gleba at maturity dingy cinnamon to buffy tan; spores ± honey brown under microscope; growing in grassy areas such as lawns . . . . (p. 250) *E. agaricoides*
1. Gleba at maturity chocolate black; fruit body short-stipitate; gleba seldom exposed at maturity; growing in mountain conifer forests . . . . (p. 251) *E. depressum*

---

195 Endoptychum agaricoides

Field identification marks. (1) Fruit bodies 3–10 cm wide, 4–12 cm high, globose to eggshaped and attached by a cord (rhizomorph) at the base; (2) in longitudinal section showing an undeveloped "stalk" (called a columella), a spore bearing area (gleba) and a "peridium" (homologous to the cap of a mushroom that has not expanded); (3) color white at first becoming buff to dingy tan or pale leather color; (4) gleba white at first but by maturity pale brown and ± powdery.

Edibility. Edible when young and the gleba is soft and white. Those who have reported to me on its quality compare it with the other large puffballs.

When and where to find it. Often common in arid regions such as our Southwest after wet weather, but it is also found on lawns in the Rocky Mountain area. It may appear late in the spring, in the summer after showers, or in the fall.

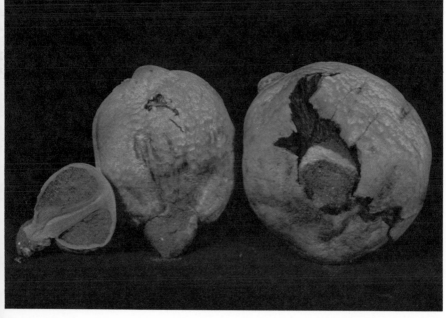

195 *Endoptychum agaricoides*   *About one-half natural size*

Microscopic characters. **Spores** 6–8(9) x 5.5–7 μ, smooth, ± thick-walled, dingy buff under the microscope. **Cystidia** none. **Basidia** 4-spored.

Endoptychum depressum 196

Field identification marks. (1) The chocolate to blackish gleba ± powdery in old age; (2) the broadly convex fruit body with a short, thick stalk; (3) flesh when fresh staining yellow where injured; (4) old fruit bodies with a ± disagreeable odor.

Observations. The fruit bodies of this mushroom are apt to be mistaken for young ones of a species of *Agaricus* — the gleba becomes chocolate black, for instance, as in *Agaricus edulis.*

Edibility. I have no reports on it. It has probably been eaten by some people who thought they had young material of a species of *Agaricus.*

When and where to find it. It has been found most frequently in the Salmon River drainage of Idaho, often under aspen or lodgepole pine. Present evidence indicates that it is not a mycorrhiza-former, or if so is of the generalized type. Fresh material has been found in July around McCall, Idaho. Since the old fruit bodies persist in the duff later records do not necessarily indicate late fruiting.

Microscopic characters. **Spores** 6–9(10) x 5–8 μ, smooth, globose to subglobose, lacking a pore, wall colored (in KOH) as in *Agaricus* (dark chocolate color). **Basidia** 4-spored (rarely 1- or 2-spored). **Cystidia** none. **Subhymenium** cellular. **Clamps** none.

196 *Endoptychum depressum*

About one-half natural size

197 Lycoperdon perlatum

Field identification marks. (1) The cone-shaped spines over the slightly immature fruit body (they fall off later leaving ± round spots); (2) sterile base (area below the gleba) chambered; (3) spore mass at maturity olive brown to ± date brown; (4) growing on or near accumulations of organic debris.

Edibility. Edible in young state (when gleba is pure white). Do not use specimens in which the gleba has started to turn yellow. It is rated as one of the best of the smaller puffballs.

When and where to find it. It is found most abundantly along roads where woody materials were used as fill and covered with dirt, but *I do not* recommend using such fruit bodies for food, especially if the roadside has been sprayed. It is, basically, a lignicolous fungus and can be collected in quantity along old rotting logs throughout the area during late summer and fall.

Microscopic characters. **Capillitium** of simple or sparingly branched threads 5–6(9) μ wide, with brown thickened walls, ± flexuous, walls pitted. **Spores** globose, 3.2–4.5 μ, smooth to very minutely warted.

198 Macowanites americanus

Field identification marks. (1) The color of the cap is typically olive, pink, lilac, purplish to purple red — usually mixed on one cap or present in various combinations; (2) the gleba ("aborted" gills) are ochraceous when mature; (3) the cap surface soon cracking into large flakes (see photo); (4) the "gills" very convoluted and almost unrecognizable as such; (5) the fruit body with the appearance of a small aborted *Russula;* (6) spores with amyloid ornamentation; (7) often found in clusters.

Observations. The species in this genus are recognized on microscopic features for the most part so only the most common one is included here. Species of *Macowanites* are like those of *Russula* in all respects save that the spores are not discharged from the basidia, and that the gills (gleba) are so organized into chambers that most spores could not fall free if they were discharged. More species of *Macowanites* are known for the Pacific Northwest than for any other part of the world.

Edibility. According to mushroom hunters in central Idaho, this species is edible. All the species of this genus are pre-

197 *Lycoperdon perlatum*   *About natural size*

198 *Macowanites americanus*   *Slightly less than natural size*

ferred by rodents to the extent that during the height of the season in late July one often finds whole clusters represented only by the bases of the stalks.

When and where to find it.   The species is most abundant in the Salmon River country of Idaho where it can be collected regularly after summer rains, the peak of the season being late July or early August. The genus, however, is not confined to the semiarid areas, as one species can be collected quite regularly at Cape Lookout State Park on the Oregon Coast. In Idaho they are found mostly in the spruce-fir zone in the mountains.

Microscopic characters. **Spores** 8.5–13.5 x 8–12 μ, subglobose to short ellipsoid, yellowish; ornamentation strongly amyloid, consisting of warts and short spines partly connected to form broken lines or a broken reticulum, ornamentation 0.5–0.8 μ high; plage area with a strongly amyloid area of amorphous material. **Basidia** 4-spored or more rarely 1-, 2- or 3-spored. **Pseudocystidia** 40–90 x 7–12 μ. **Epicutis of peridium** of projecting pseudocystidia becoming decumbent finally and arising from the layer of subgelatinous hyphae forming the cuticle or arising below it and projecting through it. **Clamps** absent.

Astreus

Key to Species

1. Expanded fruit body 6–15 cm broad; upper surface of rays conspicuously transversely cracked; spore sac rupturing irregularly . . . . (p. 254) *A. pteridis*
1. Expanded fruit body 2–5 cm broad; surface of rays merely rimulose; spore sac opening by a pore . . . . . .
. . . . (p. 255) *A. hygrometricus*

---

199   Astreus pteridis

199 *Astreus pteridis*   *About two-thirds natural size*

Field identification marks. (1) The fruit body opens in the manner of an earth star; (2) the large size (when expanded up to 15 cm wide); (3) outer shell 3–6 mm thick, woody when dry; (4) inner layer of rays conspicuously cross-checked; (5) spore sac 2.5–4 cm wide and globose; (6) spore sac opening by rupturing.

Edibility. Not edible because of its consistency.

When and where to find it. Solitary or in small groups on soil and humus in the fall, often in old logging roads in stands of Douglas fir but not limited to a single tree species, not abundant but a few fruit bodies are usually found each season; Pacific Northwest mainly but it has been found in Iowa also.

Microscopic characters. **Capillitium** of long hyaline thick-walled threads 5–7.5 μ wide. **Spores** large, 8–11.5(12) μ, globose, obscurely roughened.

Astreus hygrometricus 200
(False Earth Star)

Field identification marks. (1) The fruit body has an outer wall opening out and separating into segments arranged in an approximately starlike manner; (2) the spore case is roughened and opens by a pore at the apex; (3) upper surface of the segments of the outer wall (rays) are areolate, and the rays close back over the spore case in drying out, but open out again when wet; (4) the cocoa-colored spore mass (gleba).

Observations. This species and a few others resemble the true earth stars *(Geastrum)* so closely that they were placed

200 *Astreus hygrometricus*   *About one-half natural size*

in that genus by early authors. However, according to Morgan who described *Astreus,* it differs from *Geastrum* in lacking open chambers in the young gleba and not having a true hymenium. Hygroscopic species of *Geastrum* are abundant in arid regions of the West.

Edibility. Not edible because of its consistency.

When and where to find it. Scattered to gregarious on sand dunes, beaches and waste sandy soil, very common and widespread, but apparently not common in the West. Freshly developing fruit bodies are most likely to be found in the spring.

Microscopic characters. **Gleba** white at first, cocoa brown at maturity. **Spores** 7–10.5 μ, globose, with a very thin hyaline envelope, interior to this is a thickened colored wall with many pores (both round and irregular) so that the outer surface of the colored wall does not appear entirely smooth. **Capillitium** as yellowish thick-walled cells with incrusting material (as in *Geastrum,* in part, but as a rule more branched).

Nidula (Bird's Nest Fungi)

201   Nidula candida

Field identification marks. (1) The urn-shaped fruit body with small pill-like structures contained within it; (2) a thin layer of tissue over the top of the fruit body which at first covers the "pills" (peridioles); (3) the dull cinnamon colored scurf over the sides and base of the fruit body; (4) the gel-

201 *Nidula candida*   *Slightly less than natural size*

like substance in which the peridioles are embedded. These can be seen best soon after the covering tissue of the fruit body (the epiphragm) breaks.

Observations. Under the above name in *Mushrooms in their Natural Habitats,* p. 30, a fungus was illustrated which Palmer subsequently identified as *Nidula niveo-tomentosa.* The fruit bodies illustrated in the present work are the same as specimens identified by Palmer as *Nidula candida.* Peck's original description calls for a snow white fungus growing on the ground. Because of the obvious discrepancies to be noted in the photograph as compared with the original description, I wish to point out certain facts: (1) The fruit bodies have a dull cinnamon scurf over the basal part of the structure (the "nest") and over the epiphragm when in growing condition; (2) the "nest" is at first filled with a gel; (3) the peridioles (the "eggs" or "pills" in the nest) are light brownish; (4) it is not uncommon for a new fruit body to develop within an old one; (5) the species grows on old berry canes (species of *Rubus)* and other woody material.

Each peridiole in the nest contains a number of basidiospores, and it is the peridioles which become scattered about by rain or by animals. Because spores of the correct mating types are contained in each peridiole, upon germination a mycelium termed a dicaryon is produced directly even though large numbers of spores are not produced as in nearly all the other basidiomycetes described in this work.

Edibility. The fruit bodies are too small and too woody to attract the mycophagist whether he be a man or a worm.

When and where to find it. It is a very common species late in the fall in the West in dense thickets of berry canes along streams or in the fog belt along the coast. The fruit bodies may persist for a year at least and, I suspect, even longer.

Microscopic characters. **Peridioles** pill-like in shape, 1–2 mm wide, smooth, brownish to pallid. **Spores** 6–10 x 4–8 μ, globose to ellipsoid, hyaline. **Clamps** present on the hyphae surrounding the peridioles.

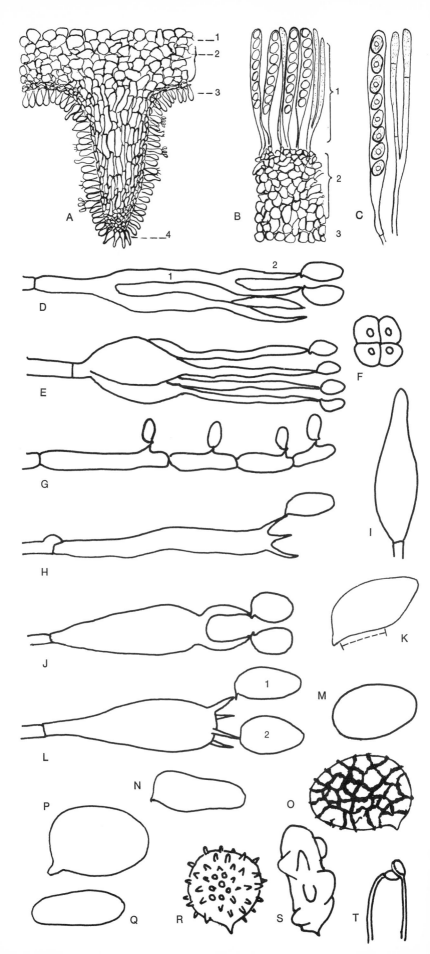

Figs. A–T. Microscopic characters

Microscopic Characters

These figures illustrate a few of the characters basic to the study of fleshy fungi. They are not necessarily illustrated by species described in the text.

Figure A represents a cross section through the cap and gill of a typical "mushroom" or gilled fungus. A-1 shows a cuticular region not distinct as a layer from the cap trama A-2. The cap trama continues down into the gill where it is termed the gill trama and in many fungi is distinguished from the cap trama mainly by a somewhat vertical interwoven arrangement of the hyphae. A-3 represents the hymenium composed of a palisade of basidia. A-4 represents cystidia on the gill edge (termed *cheilocystidia,* those on the side are termed *pleurocystidia).* Both terms denote position, not *type* of cystidium.

Figure B is a semidiagrammatic section of a piece of a fruit body of a cup-fungus to show asci with spores and paraphyses (B-1) making up the hymenium. It is readily seen how the hymenium in a Basidiomycete and an Ascomycete differ. B-2 is the hypothecium or tissue under the hymenium in a cup-fungus. B-3 represents a simple type of *exciple* not readily distinct from the hypothecium. Figure C is of an *ascus* and a *paraphysis* shown isolated from the hymenium.

Figure D represents the basidium in the Dacrymycetales: D-1 is the mature *metabasidium,* D-2 the mature *epibasidium.* In this group the young basidium is termed a *probasidium.* The terms are likely to be confusing to the beginner because they merely indicate different names for the parts of the maturing basidium.

Figure E represents the basidium in the Tremellales. The young stage (probasidium) is a single cell, but as it matures two walls form which are oblique or roughly parallel to the long axis of the cell, thus producing a 4-celled apparatus (shown in figure F as a cross section). From the apex of each of these cells an elongated tube extends to the surface of the hymenium (remember that the fruit body is jellylike in consistency) and the basidiospores are formed at the tips.

Figure G represents the basidium of the Auriculariales, and, as one will note, it does not resemble the other basidia in any way, being essentially a septate hypha in which each cell produces a spore. It is a basidium because the spores are those of the sexual stage of the fungus.

Figures H, J, and L represent the most common type of basidium, a single cell cut off from the parent hypha by a cross wall at the base. The shape does not vary much in the thousands of species featuring them, but the greatly elongate type is a feature of the Cantharellales, the type with the large bowed sterigmata is found in a number of groups of fleshy fungi but is one of the generic characters of *Clavulina* in the Clavariaceae. The sterigmata (the prongs on which the spores form) are often very fine — much more than as shown in species number 3.

Figure I illustrates the most common type of cystidium in the hymenium of Basidiomycetes and is described as *fusoid-ventricose.* Most cystidia are variants of this basic type.

Spore shapes are illustrated by figures K, M, N, P, and Q. Figure K represents a profile or side view of the same spore shown by M (in face view). In other words spores often show bilateral symmetry. In K, the area indicated by the dotted line is known as the *suprahilar depression.* The shape of such a spore as seen in profile is termed *inequilateral.* The view shown by M (face view) is described as *ovate* (egg-shaped). Figure N shows a spore described as elongate-inequilateral or "boletoid" because so many boletes have spores of approximately this shape. Figure P, not counting the apiculus, is described as elliptic, or ellipsoid (if it is the same shape in both profile and face views). Figure Q might be described as elongate-ovate in the view shown (or sub-oblong — meaning almost oblong).

Figure O shows a spore with reticulate ornamentation; figure R shows an echinulate spore; and figure S a nodulose spore.

Figure T illustrates an ascus with an operculum or lid, a feature requiring high magnification under the microscope to view it properly. Many Discomycetes have this type of ascus.

Glossary

ABNORMAL (of a specimen): not properly developed. Used to describe a difference which is very pronounced but not inherited, such as the development of gills on top of a cap in a gill mushroom.

ABRUPT: terminating suddenly or sharply differentiated. Used to describe the base of a stalk or the apex of the bulb.

ACICULAR: needle-shaped.

ACRID (taste of a raw mushroom): causing a biting or pricking sensation on the end of the tongue.

ACUTE: pointed; (of gills) sharp-edged.

ADNATE (of gills): bluntly attached to the stalk.

ADNEXED (of gills): broadly notched at the attachment to the stalk.

AGARIC: a sporocarp (fruit body) of a gilled mushroom (used as a short term for a member of the *Agaricales).*

AGGLUTINATED: stuck together.

ALUTACEOUS: more or less the color of buckskin (dull yellow brown).

ALVEOLATE (of a surface): with shallow broad pits.

AMYLOID (of spores): staining violet, blue or bluish gray in Melzer's solution (an iodine solution).

ANNULUS: the ring on the stalk left by the breaking of the inner veil.

APEX (pl. apices): the tip of the part described.

APICAL (of a stalk): the part of the stalk near the line formed by the attached gills — generally the uppermost part of the stalk.

APICULATE (of a spore): with a slight narrow basal projection typically oblique.

APOTHECIUM: the cuplike or saucerlike fruit body of some *Ascomycetes.*

APPENDICULATE (of the cap margin): with pieces of the veil hanging along the margin.

APPRESSED (of fibrils or hairs on the cap): lying flat on the surface.

AREOLATE: cracked in the pattern of a mudflat as it dried out.

ASCI (sing. ascus): the cell *in* which reduction division takes place in the *Ascomycetes* and in which the spores of the sexual stage are produced.

ASCOSPORE: the spore produced in an ascus.

ASPERULATE: very slightly roughened.

AURIFORM: ear-shaped.

AZONATE (of surface of a cap): lacking concentric bands of different color.

BASIDIOCARP: a fruit body which bears basidia.

BASIDIOLE: an immature basidium.

BASIDIUM (pl. basidia): the cell in a mushroom in which reduction division takes place and *on* which the spores of the sexual state are formed.

BOLETINOID (of the pore surface in the *Boletaceae):* having pores elongated radially to the degree that their pattern is intermediate between round pores and gills.

BOLETOID: having essentially circular pores on the hymenophore; also used in the sense of *Boletus*-like for the fruit body; of spores, like those typical of a *Boletus* — elongate-inequilateral.

BROAD (of gills): a relative term to describe the depth of

the gills. It is contrasted with "moderately broad" and "narrow."

BUFF (a color): a pale yellow toned with gray, that is, a dingy pale yellow.

BULBOUS (of a stalk): having an oval to abrupt enlargement (bulb) at the base.

CAESIOUS: pale grayish blue.

CAMPANULATE (of a cap): bell-shaped.

CANESCENT: having a bloom, or appearing as if coated with a light hoarfrost.

CAP: the umbrella-like expansion on the apex of the stalk in a mushroom. It bears the gills, teeth, or pores on the underside. "Pileus" is the technical term applied to it.

CAPILLITIUM: the specialized hyphae mixed with the spores in (typically) members of the true puffballs.

CARMINOPHILOUS GRANULES: granules in the basidium which stain darkly in aceto-carmine stain.

CAULOCYSTIDIA: sterile cells occurring on the surface of the stalk.

CELLS (of fungi): the living protoplasmic units into which the hyphae are divided.

CESPITOSE (also caespitose): occurring in clusters.

CHEILOCYSTIDIA: sterile cells occurring on the edges of the gills or pores in an agaric or bolete.

CINEREOUS: pale gray (color of wood ashes).

CLAMPS (clamp connections): a short looping branch on a fungous hypha originating at the distal cell of a pair of cells and fusing with the adjacent cell near the apex of the latter.

CLAVATE: club-shaped, of a fruit body broad at apex and narrowed to the base; but in a stalk the order is reversed — it is narrow at the apex and broad at the base.

CLAVATE-MUCRONATE: a clavate cell (broad at apex and narrowed to the base) and with an apical beak.

CLOSE (of gills): a relative term to indicate the spacing of the gills.

CLUB: often applied to single fruiting bodies of the coral fungi in which the upper part is enlarged.

COLLAR: a close-fitting roll of universal-veil tissue around the apex of the bulb in an *Amanita*.

COLUMELLA: a column or vein of sterile tissue extending through the spore-bearing tissue; used mostly in gastromycetes where in secotioid species the upper part of the stalk is termed the columella or the whole a "stipe-columella."

CONCHATE: the shape of a seashell.

CONCOLOR: the same color overall (or in comparing parts of a fruit body meaning "all of the same color").

CONCOLOROUS: used with "with" — meaning the same color as the other part mentioned, such as "cap concolorous with the stalk."

CONCRESCENT (of the caps): grown together at their margins to form a compound cap.

CONFIGURATION: relative disposition of the parts on an object or the form which this produces.

CONIDIAL: applied to the spore state for nonsexually produced spores, i.e., the "conidial state."

CONIFER: a cone-bearing tree, such as a pine or fir.

CONSISTENCY: the firmness, density or solidity of the tissue which makes up the fruiting body.

CONTEXT: the flesh of the cap and stalk (regardless of whether it is soft, tough, or woody).

CRISPED (of gills): crinkled.

CRISTATE (of spore ornamentation): with ornamentation as small crests or ridges.

CROCEOUS: a rich yellow, like a crocus.

CROWDED (of gills): spaced very close together. "Crowded," "close," subdistant," and "distant" are the four relative terms used to describe gill spacing.

CUTICLE (of a cap): the differentiated surface layer. (Not all species have such a layer.)

CUTIS: the surface covering. It applies to either cap or stalk.

CYSTIDIA: sterile cells which occur scattered in the hymenium among the basidia and at maturity differ in shape from the basidia.

CYSTIDIOID: shaped like a cystidium (typically ventricose in the midportion and tapered to the apex).

DECURRENT (of gills): extending downward on the stalk.

DEPRESSED (of a cap): the central part sunken slightly below the margin.

DEXTRINOID: giving a reddish brown to vinaceous red color in Melzer's solution.

DICHOTOMOUS: forked.

DISC: the area of the cap over the center as contrasted with the marginal area.

DISSEPIMENTS: the walls of the tubes in a polypore.

DISTANT (of gills): spaced far apart.

DUFF: the accumulation of organic material on and in the soil of a forest.

ECCENTRIC: off center.

ECHINULATE: furnished with fine needlelike projections (used most in describing spore ornamentation).

ECTAL (excipulum): the outer layer of the "cup" in a *Discomycete.*

ELLIPTIC: ellipsoid, in the form of an ellipse, rounded at both ends and sides curved outward (contrasted to "oblong" in which the sides are parallel).

EMARGINATE (of gills): notched at the attachment to the stalk.

ENDOPERIDIUM (of a puffball): the inner wall (usually forming the spore sac).

ENTIRE (of gill edges): even, in contrast to "serrate," etc., in which the edge is cut into small teeth or broken up in some other manner.

ENZYME: an organic compound capable of causing changes in other compounds by catalytic action.

EPICUTIS: the surface layer of the cap or stalk in a fruit body.

EPISPORE: the outermost layer of the spore wall.

ERODED (of gill edge): uneven in an irregular way.

ESCULENT (as used in this work): an edible fungus.

EVEN (of cap surface): with no depressions or elevations.

EXCENTRIC: with the stalk not centrally attached to the cap.

EXCIPULUM: the wall of the apothecium in a *Discomycete* (it is often layered).

EXOPERIDIUM (of a puffball): the outer layer of the spore sac wall (often in the form of granules, warts or flakes).

FAIRY RING: a naturally occurring circle of fruiting bodies of any mushroom.

FALSE VEIL (in the *Boletaceae*): a tissue that grows out from the margin of the cap but does not become intergrown with the stalk, though the remains of the veil may form a ring around it.

FARINACEOUS (of odor and taste): like that of fresh meal.

FeSO₄: the symbol for a variety of iron salts in solution in water; the usual concentration is about 10 percent. A positive reaction on the tissue of the basidiocarp is green to olive to olive gray or in some fungi pinkish. A strong reaction is olive black.

FIBRILLOSE (of a cap or stalk): covered with appressed hairs or threads (fibrils) more or less evenly disposed.

FIBROUS (flesh of stalk): composed of tough stringy tissue.

FLESH (of a mushroom): the tissue of the cap.

FLESH COLOR (a color): pinkish to the color of raw meat.

FLESHY: soft in consistency, decaying readily, contrasted with woody or membranaceous.

FLOCCOSE SCALY (of a cap): having tufts of woolly material, usually remnants of a universal veil.

FREE (of gills): not attached to the stalk at any time during their development.

FRUCTIFICATION: same as fruit body, and can be applied to an ascocarp or a basidiocarp.

FRUITING BODY: the part of the fungous plant developed for the purpose of producing and liberating spores. "Basidiocarp" is another term applied to it. I have used the term "mushroom" to mean the same thing.

FURFURACEOUS: roughened by minute particles.

FUSCOUS (a color): the color of a storm cloud to a dark smoky brown, considerable violet evident, but the amount varies.

FUSOID-VENTRICOSE: broadly spindle-shaped.

GENUS: a grouping of species. It is the first main group above the rank of species and its name is given ahead of the epithet of the species in the formal Latin name of the species, i.e., *Cortinarius violaceous.* There are over 800 kinds of *Cortinarius* in North America, but only one can be properly called *Cortinarius violaceous.*

GILLS: the knifeblade-like radially arranged plates of tissue on the underside of a mushroom cap. "Lamellae" is the technical term.

GLABRESCENT: becoming bald.

GLABROUS: bald (lacking veil remnants or fibrils on the cap) as applied to mushrooms.

GLANDULAE: aggregations of resin-secreting caulocystidia.

GLANDULAR DOTTED (of a stalk): with aggregations of caulocystidia which secrete a resinous material — their appearance on the stalk varies from broad irregular smears to fine dots darker in color than the ground color of the stalk.

GLEBA: the spore mass in a puffball (or related fungi); it may be dry or slimy and often fills most of the interior of the fruit body.

GLUTINOUS: covered with a slimy to sticky layer.

GRANULOSE: covered with granules, either free or attached.

GREGARIOUS: growing close together but not in clusters.

HEAD: a more or less globose enlargement at the apex of the stalk in some fungi; the term contrasts with "cap," which is an umbrella-like expansion.

HELVELLOID: shaped as the cap in *Helvella,* roughly saddle-shaped.

HETEROMEROUS (of the tissue of the fruit body): composed of filamentous hyphae and groups of vesiculose cells.

HOLOTYPE: the collection designated the type in the formal Latin diagnosis of a new species or variety.

HOMOGENEOUS (of the cap): of the same consistency throughout — lacking distinct layering.

HUMUS: the fine particles of organic material in the soil. The term "duff" applies to coarser material.

HYALINE: colorless, transparent.

HYGROPHANOUS (of the cap): the feature of being one color when moist and typically a sharply different often paler color when moisture has escaped.

HYMENIFORM: with cells arranged as in a hymenium.

HYMENIUM: with the spore bearing cells arranged in a palisade.

HYMENOPHORE: the structure bearing the hymenium. The gills in a mushroom, pores in a bolete, or the teeth in a *Hydnum,* etc.

HYPHA (pl. hyphae): the basic unit of structure of the spawn as well as of the fruit body of the mushroom plant. It is a thread, and in both the Ascomycetes and Basidiomycetes the threads are partitioned by cross walls (septa).

HYPOGEOUS (of fungi): with fruit bodies developing and maturing in the ground in contrast to *epigeous* where they come up out of the ground as in most mushrooms.

HYPOTHECIUM: the region in a cup-fungus fruit body below the hymenium.

IMBRICATE (of scales): overlapping one another like shingles on a roof.

INAMYLOID: not giving a violet, blue or blue gray reaction in Melzer's solution. The object tested remains hyaline to yellow. See dextrinoid also.

INEQUILATERAL (of spores): subfusiform — if the spore is viewed in profile the dorsal hump or convexity and the ventral bulge are not exactly one above the other — the dorsal hump usually being toward the base and the ventral one toward the apex (slightly).

INNATELY (of fibrils): with the bases of the fibrils attached to the tissue of the cap (not separate veil remnants).

INTERVENOSE (of gills): with cross veins.

ISABELLINE: a pale olive tan or dingy olive ochraceous color.

IXOCUTIS: a surface layer of gelatinized hyphae.

IXOTRICHODERM: a layer of upright hyphal projections in a slime matrix (a turf).

KOH: the symbol for potassium hydroxide, I use a 2.5 percent to 3 percent aqueous solution. It can be applied directly to the fruit body for ascertaining color changes, or used as a mounting medium to revive material and at the same time to ascertain color changes on spore and hyphal walls as well as pigment deposits.

LAMELLAE (gills): the plate-like structures forming the hymenophore of a mushroom.

LAMELLATE: having gills.

LAMELLULAE: the gills that extend only part way from the cap margin to the stalk.

LATERAL: attached by one side.

LATEX: a juice, usually milky, but colored in some mushrooms, which is exuded when the plant is injured.

LATICIFEROUS (of hyphae): containing a latex (milklike or variously colored).

LIGNICOLOUS: characteristically found growing on wood.

LIGNIN: one of the main constituents of wood.

MACROCYSTIDIA (in *Russula*): large pointed projecting cystidia with a content of globules.

MARGIN (of a cap): the outermost part of the cap, near the edge and including it.

MARGINATE (of gills): with edges differently colored than the faces.

MEDULLARY EXCIPULUM: the layer interior to the outer layer of the wall of an apothecium.

MELZER'S REAGENT: an iodine solution as follows :

KI	1.5 gms
Iodine	0.5 gms
Chloral hydrate	20.0 gms
Water	22.0 gms

MILD (of taste): lacking any distinctive taste, bland.

MONOMITIC (of hyphal systems): with only one type of hypha.

MONOTYPIC (of a genus): containing only the type species.

MULTIPLEX: applied to an extremely compound sporocarp.

MYCELIOID: fluffy in consistency with the fluff consisting of hyphae. Also used in the sense of being covered with mycelium.

MYCELIUM: the collective term for all the threads forming the vegetative phase of an individual fungous plant.

MYCORRHIZA: the combination of the tree rootlet and the fungus mycelium surrounding or invading it.

NAKED: lacking a veil or other covering such as appressed fibrils or pruina.

NARROW (of gills): a relative term, the opposite of broad, indicating depth of gills.

OBTUSE: blunt, not pointed.

OBVENTRICOSE (of a cystidium): with the swelling distal to the midpoint of the cell.

OLEIFEROUS (hyphae): hyphae in which waste products are stored and which as a result have a content different in appearance from the other hyphae of the tissue as viewed under a microscope. See laticiferous hyphae also.

OPERCULATE: having a lid (more or less as on a coffee pot). Found on the ascus of some *Discomycetes*.

ORNAMENTATION (of spores): any discontinuity in or on the wall. The pattern is very constant from species to species and furnishes excellent taxonomic characters for species recognition, but the spores of many species remain smooth.

OVATE: the shape of a longitudinal section through a chicken egg.

PAPILLA: a small projection on a hyphal wall (nipplelike).

PAPILLATE: having a papilla.

PARAPHYSES: as used here it applies to the sterile threads in the hymenium of cup-fungi. They are often highly colored.

PARASITE: one organism residing on a second one from which it extracts sustenance.

PARTIAL VEIL: the inner veil which extends from the margin of the cap to the stalk and at first covers the gills (or pores). Contrasted to universal veil.

PEDICEL: the narrowed basal part of a cell, basidium, cystidium, ascus, etc.

PEDICELLATE: with a pedicel.

PELLICLE: a layer of narrow, appressed gelatinized hyphae (or of hyphae which secrete a mucous).

PENDANT: hanging down.

PERIDIOLE (in bird's nest fungi): the small pill-like structures in the "nest" consisting of a wall around a group of spores.

PERIDIUM (of puffballs): the wall of the spore sac (often layered).

PERITHECIA (of *Ascomycetes*): flask-shaped structures opening by a pore at the apex of a neck and in which asci develop. The spores are finally forced through the neck and out through the pore.

PICRIC YELLOW: the color of picric acid (an intense yellow).

PILEATE: having a cap.

PILEOCYSTIDIA: cystidia on the cap of a sporocarp.

PILEOLI: small pilei (caps).

PILEUS (pl. pilei): the cap of the mushroom, bolete, polypore, or *Hydnum,* or any similar structure which supports the hymenophore on its underside — and a term often used for any broadly expanded stalk apex.

PLAGE (of basidiospores): the depressed area on the ventral face of the spore just distal to the apiculus, often distinctly depressed (when it is then termed the "suprahilar depression").

PLANO-CONVEX (of the cap): nearly flat but the center slightly higher.

PLEUROCYSTIDIA: cystidia on the face of the gill, tube, or spine.

PLEUROTOID: with the habit of having the cap attached to the substratum at its base, or stalk excentric to lateral.

PLIANT: flexible.

PLICATE: folded, plaited, like a fan.

PORES: the minute to distinct holes in the layer of tissue on the underside of the cap. They contrast with the gills in a gill fungus.

POROID: having pores.

PRUINA: a faintly powdery appearance caused by presence of caulocystidia or pileocystidia.

PSEUDOCYSTIDIA: cystidia with oily to resinous content. The term is found most frequently in the descriptions of species of *Russula* (see macrocystidia also, which are a type of pseudocystidium).

PSEUDOPARENCHYMATOUS (of fungous tissue): cellular in the sense of more or less isodiametric cells but basically hyphal in origin.

PSEUDORHIZA: the term to designate the rootlike process often extending deep into the ground (actually it grows up to the surface; the fruit body forms on it).

RAPHANOID (of taste): sharp like a radish.

REDUCTION DIVISION (of nuclei): those divisions which result in reducing the number of chromosomes by half in the daughter nuclei, also known as the process of meiosis in contrast to mitosis in which the chromosomes split in half with an equal number of halves going to form each daughter nucleus.

REFRINGENT (of hyphae): with a different index of refraction than for normal hyphae *(refractive* is synonymous).

RETICULATE: a pattern of connected lines resembling somewhat that of a woven wire fence.

RHIZOMORPH: a stringlike or thin ropelike aggregation of hyphae — part of the spawn of the mushroom plant.

RUGOSE (of the cap): wrinkled variously.

SCALES (of cap or stalk): torn portions of cuticle or veil. Usually these remnants are in some sort of pattern.

SCLEROTIUM: a tuberlike small body of fungous tissue often

rather hard in consistency and usually capable of "germi-nating" to give rise to a fruit body.

SCLEROTOID: resembling a sclerotium.

SCROBICULATE (of the stalk in *Lactarius):* with large shiny depressed spots or areas.

SCRUPOSE (of a cap surface): very coarsely roughened.

SECOTIOID: resembling the fruit body of species of *Secotium* (basically that of an unexpanded mushroom fruit body).

SESSILE: attached directly to the substrate, lacking a stalk.

SHEATH: boot, usually veil remnants on the lower part of the stalk. This condition is called peronate.

SIMPLE (of fruiting bodies): unbranched.

SLIMY: covered with a viscous material.

SMOOTH (of a surface): even, lacking wrinkles or projec-tions.

SOLID (of stalk): lacking a central cavity and similar in tex-ture throughout.

SORDID: dirty or dingy in appearance.

SPECIES: a species is one kind of organism.

SPHAEROCYSTS: bubble-shaped hyphal cells (also termed *sphaerocytes* by some).

SPONGY (of flesh): soft and tending to be water-soaked.

SPORE DEPOSIT: a mass of spores deposited naturally (or from a mushroom set up so that it will shed spores), which is visible to the naked eye.

SPOROCARP: any fungous fruiting body bearing spores.

SQUAMULOSE: covered with small scales.

STERIGMATA: the prongs of the basidium on the apex of which the basidiospores are formed.

STIPE: the stalk of the fruit body.

STIPITATE: having a stalk.

STIRPS (pl. stirpes): an unofficial category indicating rela-tionship. A stirps consists of a central species with "satel-lite" species grouped with it. The name of the central species is the source of the name for the stirps (i.e., stirps *Violaceus* — for *Cortinarius violaceus)* and closely related species.

STRIAE: the radiating lines or furrows on a mushroom cap, or the longitudinal lines on a stalk.

STRIATE: having radiating lines or furrows.

STRIGOSE (of base of stalk): having coarse long hairs.

SUB-: a prefix meaning almost, somewhat, or under.

SUBCUTIS: the layer in the cuticular complex of the cap which is next to the context proper (the inner-most dif-ferentiated layer of the cap cuticle).

SUBDECURRENT: slightly decurrent.

SUBDISTANT (of gills): between close and distant.

SUBICULUM: the mat of hyphae on which a structure is produced.

SUBOPERCULATE: applied to asci in which a lid on the ascus is not as characteristic as in the operculate condition, or to an intermediate stage between operculate and in-operculate.

SUBRAPHANOID: with a sharp taste somewhat like that of radish.

SUBSTRATE: the material to which a fruit body is attached.

SUBUMBONATE: with a slight hump in the center of the cap.

SULCATE: rather deeply grooved but not plicate.

SUPERFICIAL: merely resting on the surface, not attached.

SUPERIOR (of a ring): attached above the middle of the stalk.

SUPRAHILAR DEPRESSION: see plage.

SYNONYM: one of the names that has been given to a spe-cies but which is not the correct (valid) name.

TAWNY (a color): about the color of a lion.

TAXA (sing. taxon): a word to designate a category or categories of plants from *form, variety, subspecies, species, series, section, subgenus, genus,* etc. Each is a taxon.

TERETE: circular (in section), used to describe the stalk.

TERRESTRIAL: growing on the ground (in contrast to "lignicolous").

TOMENTUM: a covering of soft hairs.

TRAMAL PLATE: the tissue (of a gill, for instance) between the hymenial layers. In a puffball the tramal plates break down as the gleba matures.

TRICHODERM(IUM): an upright to tangled layer of hyphal ends (2–6 cells long) usually over the cap. A hymeniform layer is one cell deep.

TRUNCATE: chopped off (the apex flattened).

TUBERCULATE: with low bumps or irregularities (a type of spore ornamentation).

TURBINATE: top-shaped.

TURF: a tangled layer of narrow hyphae nearly erect to semidecumbent (intergrading with a trichodermium).

TWO-SPORED BASIDIA: basidia typically bear four spores but in certain species the basidia bear only two spores.

UMBER (a color): tobacco brown or darker.

UMBO (of the cap): with a hump over the center.

UMBONATE: furnished with an umbo.

UNCINATE: the gill having a narrow sinus at the junction with the stalk.

UNGULATE: hoof-shaped.

UNIGUTTULATE: with one droplet (ascospores often have 1–3 droplets in them).

UNIVERSAL VEIL: the veil which envelops the young fruiting body in some genera and species. It is an outer layer of tissue distinct from the cap and stalk.

URCEOLATE: urn-shaped.

VARIANT: an unofficial designation for a collection differing slightly from the type form but for which the user of the term does not wish to use a formal designation. One may speak of variants of a species or of a variety.

VEIL: a layer of tissue covering the young developing hymenium.

VENOSE: with veins.

VENTRICOSE: a swelling (used mainly in connection with hyphal cells).

VERRUCOSE: warty.

VERSIFORM (of such cells as cystidia): of many forms.

VESICULOSE: bubble-shaped.

VINACEOUS (a color): the color of red wine or a paler red.

VIRGATE: streaked.

VISCID: sticky to the touch.

VOLVA: the remains of the universal veil left around the base of the mushroom after the veil has broken.

WARTS (on a cap): small chunks of universal-veil tissue.

ZONATE: marked with concentric bands (zones) of a different color than the remainder of the surface.

Bibliography

Bigelow, Howard E. 1970. *Omphalina* in North America. *Mycologia* 62: 1–32.

Corner, E. J. H. 1950. A monograph of *Clavaria* and allied genera. *Ann. Bot. Mem.* 1:1–740.

_____. 1970. Supplement to "A monograph of *Clavaria* and allied genera." *Beihefte zur Nova Hedwigia,* 3:1–299.

Donk, M. A. 1966. *Osteina,* a new genus of the *Polyporaceae. Schweitz. Zeitschr. Pilzk.* 44:83–87.

Doty, Maxwell S. 1944. Clavaria, *the species known from Oregon and the Pacific Northwest.* Oregon State Monographs. Studies in Botany. No. 7. Corvallis, Oregon: Oregon State College Press.

Hesler, L. R., and Alexander H. Smith. 1960. Studies on *Lactarius-II.* The North American species of sections *Scrobiculus, Crocei, Theiogali,* and *Vellus. Brittonia* 12: 306–50.

_____. 1963. *North American species of* Hygrophorus. Knoxville: University of Tennessee Press.

_____. 1965. *North American species of* Crepidotus. New York: Hafner Pub. Co.

McKenny, Margaret, and D. E. Stuntz. 1971. *The savory wild mushroom.* Seattle: University of Washington Press.

McKnight, Kent H. 1969. A note on *Discina. Mycologia* 61: 614–30.

Mains, E. B. 1956. North American species of the *Geoglossaceae.* Tribe *Cudonieae. Mycologia* 48:694–710.

Marr, Currie D., and Daniel E. Stuntz. 1973. *Ramaria* of Western Washington. *Bibliotheca Mycologica,* Bd. 38, 1–232.

Martin, G. W. 1952. Revision of the North Central *Tremellales.* State University of Iowa *Stud. Nat. Hist.* 19(3):3–122.

Miller, O. K. 1964. Monograph of *Chroogomphus (Gomphideaceae). Mycologia* 56:526–49.

_____. 1972. *Mushrooms of North America.* New York: E. P. Dutton & Co.

Moser, M. 1967. In Helmut Gams' *Kleine Kryptogamenflora,* Vol. II/b2. Stuttgart: Gustav Fischer Verlag.

Murrill, W. A. 1915. *Western polypores.* New York: Published by author.

Overholts, L. O. 1953. *The* Polyporaceae *of the United States, Alaska, and Canada.* Ann Arbor: University of Michigan Press.

Pilát, Albert and O. Ušák. 1954? Mushrooms. London: Spring Books, 340 pp.

Shaffer, Robert L. 1962. The subsection *Compactae* of *Russula. Brittonia* 14:254–84.

Singer, Rolf. 1959. New and interesting species of *Basidiomycetes* VII. *Mycologia* 51:578–94.

_____. 1962. *The* Agaricales *in modern taxonomy,* Weinheim, Germany: J. Cramer.

Smith, Alexander H. 1947. *North American species of* Mycena. Ann Arbor: University of Michigan Press.

_____. 1949. *Mushrooms in their natural habitats.* Portland, Oregon: Sawyer's Inc.

_____. 1958, 1963. *The mushroom hunter's field guide.* Ann Arbor: University of Michigan Press.

_____. 1972. The North American species of *Psathyrella.* Memoirs New York Botanical Garden 24:1–633.

Smith, Alexander H., and L. R. Hesler. 1962. Studies in *Lactarius-III.* The North American species of section *Plinthogali. Brittonia* 14:369–440.

————. 1968. *The North American species of* Pholiota. New York: Hafner Pub. Co.

Smith, Alexander H., and Rolf Singer. 1964. *A monograph on the genus* Galerina Earle. New York: Hafner Pub. Co.

Smith, Alexander H., and Harry D. Thiers. 1971. *The boletes of Michigan.* Ann Arbor: University of Michigan Press.

Smith, Alexander H., and S. M. Zeller. 1966. A preliminary account of the North American species of *Rhizopogon.* Memoirs New York Botanical Gardens 14:1–177.

Smith, Helen V., and Alexander H. Smith. 1973. *How to know the non-gilled fleshy fungi.* Dubuque, Iowa: Wm. C. Brown Co.

Snell, Walter H., and Esther A. Dick. 1970. *The* Boleti *of Northeastern North America.* Lehre, Germany: J. Cramer.

Thiers, Harry D. 1966. California boletes. II. *Mycologia* 58: 815–26.

Index

Numbers in bold face are species numbers and those in italics are the pages on which the descriptions can be found.